The Best American Science Writing 2012

THE BEST AMERICAN SCIENCE WRITING

EDITORS

2000: *James Gleick*
2001: *Timothy Ferris*
2002: *Matt Ridley*
2003: *Oliver Sacks*
2004: *Dava Sobel*
2005: *Alan Lightman*
2006: *Atul Gawande*
2007: *Gina Kolata*
2008: *Sylvia Nasar*
2009: *Natalie Angier*
2010: *Jerome Groopman*
2011: *Rebecca Skloot and Floyd Skloot*

The Best American 2012

Science Writing

Editor: Michio Kaku

Series Editor: Jesse Cohen

An Imprint of HarperCollinsPublishers

FIRST EDITION

Library of Congress Cataloging-in-Publication Data has been applied for.

ISBN 978-0-06-211791-5

12 13 14 15 16   OV/RRD   10 9 8 7 6 5 4 3 2

Contents

Introduction *by Michio Kaku*

I HAVE A CONFESSION TO MAKE.

First, I am a science junkie. I hate to admit it. Some people can never pass up a chocolate cookie, or a football game. For me, I am a sucker for bright, shiny magazine covers that feature blazing galaxies, glorious exploding stars, or rampaging dinosaurs, which I find irresistible. So editing *The Best American Science Writing 2012* was like being a kid in a candy store.

Second, I confess that I am not your typical science editor. I am not a journalist at all, but a research physicist. While this may qualify me as an authority on the arcane details of quantum physics, it also means that, as editor, I have to be especially careful to be scrupulously fair to all disciplines in choosing the best articles. (I still vividly remember my sixth grade teacher explaining to me the difference between a journalist and a scientist. A journalist, he said, knows a little bit about everything, while a scientist knows a lot about a single thing. So, he concluded, the ultimate journalist knows nothing about everything, and the ultimate scientist knows everything about nothing. Hopefully, I have avoided these extremes.)

And third, I have to come clean and admit that I have a hidden agenda. President Obama has said that we must recapture "the Sputnik moment" for the next generation of young scientists. I agree. This is a matter of national urgency, since we are facing stiff competition from emerging nations eager to master the very technology that gave the West such vast wealth. Science is the engine of prosperity, and people in emerging nations often understand this better than those in developed nations. So it does not bode well for the future when U.S. students consistently score near dead last in international science and math competitions. Sadly, if this continues, it means that we might be graduating high school kids directly into the unemployment line.

Unfortunately, it is hard to recapture that "Sputnik moment" without Sputnik. The manned space program has been canceled; the heroics of our astronauts will soon be found only in dusty history books. But perhaps by promoting *The Best American Science Writing* series we are doing our small part in nurturing the Sputnik moment for the next generation.

I also confess that it was a difficult task being editor of this volume, since Jesse Cohen was weekly showering me with a flood of excellent, compelling, and worthy articles to choose from. But over the years, after reading thousands of science articles, I have developed my own well-defined sense of what constitutes a good science article.

First, the best science articles have a "takeaway factor." For example, I will be the first to admit that I love watching dinosaur specials on TV. They are fun to watch, but afterward I am often left with an empty feeling. Did I learn anything, or was it just eye candy and idle amusement? I often walk away feeling dissatisfied.

By contrast, the best articles teach us a new principle of nature. Scientific principles can be remarkably simple, yet they are all deceptively powerful, allowing us to make sense out of the quirky nature of the universe. In fact, a handful of scientific concepts guide our understanding of the entire universe. Once we learn them, they often stay with us for the rest of our lives. (For example, in grade school I learned that Ben Franklin discovered why it rained—rainstorms form when moisture-laden hot air collides with a mass of dry cold air. Decades later, I still remember that simple scientific principle whenever I watch the weather report.)

My favorite Einstein quote is "Unless a theory can be explained to a child, the theory is probably worthless." In other words, all the great theories of nature are based on simple principles, which are pictorial. Newtonian dynamics can be understood in terms of cannonballs and spheres moving in circles and ellipses, while relativity can be understood in terms of clocks, yardsticks, and trains. And the genetics of DNA, as well as string theory, can be understood in terms of the interactions of strings.

Second, the best science stories give us some deep insight into the human condition or our place in the universe. As Thomas Huxley once said, "The question of all questions for humanity, the problem which lies behind all others and is more interesting than any of them, is that of the determination of man's place in Nature and his relation to the Cosmos."

Whenever I read a story about the origin of humans, the discovery of new planets in space, or the death of stars, I gain a bit more understanding of where we fit into this great universe of ours.

Third, a good article must also have an engaging style. Science, of course, is notorious for being dense and riddled with equations and jargon. After all, mathematics is the language of nature. But in a volume such as this, we need to unleash the full arsenal of literary techniques to embellish and enliven the story, via analogies, metaphors, personal anecdotes, and humor. We need to summon all our literary skills to draw in the reader, like a detective story.

And last, for me, a great science article should also have a point of view. For example, to me an article that simply chronicles the human tragedy caused by a debilitating disease is not enough. It should also lay out a message of hope, i.e., a program that one day might lead to a cure. When I was a child, I still remember that millions of Americans lived in fear of polio, a crippling disease that doomed children to be trapped in iron lungs, constantly gasping for breath. I remember fund drives like the March of Dimes, which asked people to donate money to fight this disease. Back then, there were numerous articles chronicling the heartbreak and agony of stricken children. But these articles just talked about managing the disease. Only a tiny minority of articles took a different path and boldly asked, why not cure polio rather than lament it? This was the path pioneered by Dr. Jonas Salk, who finally vanquished this dreaded disease. The nightmarish vision of entire villages populated by legions of children trapped in iron lungs never materialized.

In editing this volume, I also chose a definite order for the various disparate subjects. I decided to start with the most familiar subject, our own body. Nothing is more intimate and vital to us, or more mysterious, than our own body. From there, I work outward to the environment that surrounds us, and then the space program, the universe, and finally to the very limits of science itself when it collides with religion and sensitive societal issues.

For example, in "Mending the Youngest Hearts," Gretchen Vogel writes about a revolution in medicine that is redefining what we mean by our own body. We are witnessing the creation of a "human body shop" in which we will be able to order new organs (e.g., livers, hearts, kidneys) much as we order automobile parts today. Sadly, ninety-one thousand Americans require an organ transplant, yet eighteen die every day for lack of an organ that never comes. In the future, we will be able to order human organs the way we order a fender.

In "An Immune System Trained to Kill Cancer," Denise Grady writes about a new way to fight one of the body's most dreaded diseases, cancer. Back in 1971, President Richard Nixon declared, with much fanfare, the War on Cancer. By simply throwing money at cancer, scientists were going to vanquish it. But as historians have noted, cancer actually won that round. In retrospect, it was futile to declare a War on Cancer when we did not know what cancer really was. Now, in the age of biotechnology, we realize that cancer is a disease of our genes, and that cancer can be defeated at the molecular and genetic level, e.g., by stimulating the body's own immune system.

One of science's great technologies is X-rays, which give us an unprecedented look deep into the living body. But like all great technologies, they can bite back at us. In "X-Rays and Unshielded Infants," Kristina Rebelo and Walt Bodganich report on the scandal that, a century after the discovery of radiation, infants are often unnecessarily exposed to large quantities of radiation.

For thousands of years, great kings and emperors have sought to extend their life span by finding the fabled Fountain of Youth. They could conquer entire kingdoms, yet were defeated by the aging image they saw every day in the mirror. For millennia, scientists did not even know what aging was. Today, scientists realize that aging is the accumulation of error at the molecular, genetic, and cellular level. In "Aging Genes," Jennifer Couzin-Frankel describes how, given the enormous commercial implications of aging, scientists may sometimes be fooled into believing their own press releases. We recall that the road to immortality is often paved with the corpses of incorrect theories and fads.

Human society exploded around ten thousand years ago, not just with the coming of agriculture, but with the domestication of the dog, horse, ox, cattle, and other animals. Civilization depended crucially on the horses we rode upon and the dogs we hunted with. Now, in the age of genomic medicine, we can finally piece together the long, winding, but intimate relationship we share with these animals. In "Taming the Wild," Evan Ratliff describes how via genetics, we can fill in many of the missing gaps in the lineage of animals and the process of domestication, including our own.

Closely related to exploring the mysteries of our body is the attempt by doctors, electrical engineers, and mathematicians to understand our brain, the mind, and even to create a mechanical one.

In "Beautiful Brains," David Dobbs explores the strange nature of the teenage brain. Any parent with kids is convinced that teenagers are aliens from another planet: stubborn, reckless, and sometimes self-destructive. But maybe it's not their fault, but the fault of the neurological wiring of their brain. Contrary to popular opinion, brain studies show that teenagers clearly understand the risks of their rebellious behavior, but they value the rewards from this risky behavior much more than adults do.

And in "Criminal Minds," Josh Fischman explores perhaps one of the most controversial areas of brain studies, the link between biology and crime. Throughout history, demagogues and bigots have tried to pin the "bad seed" label on certain minority groups, often with disastrous results. But now brain scans and twin studies are revealing links between criminal behavior and damage to specific parts of the brain, e.g., the amygdala and prefrontal cortex, that produces a loss of empathy and control. Fischman is careful not to say that "biology is destiny"; perhaps environmental and social factors caused these brain abnormalities in the first place, which in turn contributed to antisocial behavior.

This raises the question: how smart can we become, given the limitations of our skull and brain tissue? In "The Limits of Intelligence," Douglas Fox uses simple physics and neurology to determine how much brainpower can be squeezed into our skulls. Surprisingly, he finds that our human brain, frail as it is, probably has nearly the maximum amount of brainpower that can realistically be crammed into our skulls. So it is humbling to realize that sitting on our shoulders may be the most complex object devised by nature in this region of the galaxy.

But if "wetware" has definite physical limitations, what about high-tech hardware, which has no such limit? Perhaps by building mechanical minds, we can better understand our biological one. In "It's Not a Game," Jaron Lanier reminds us that one of the world's most advanced computers, IBM's Watson, also has its limitations, although it decisively defeated two humans on the game show *Jeopardy!* That victory, unfortunately, was overhyped by the media. After the victory by the computer, media pundits lamented the fact that one day our machines might stuff us in zoos, throw peanuts at us, and make us dance behind bars. But the stark reality is this: Watson was so primitive that you could not congratulate it. Watson had no self-awareness, no understanding that it had made history by defeating humans. Watson, after all, is a sophisticated adding machine. (But it is always conceivable that, many decades in the future, a machine might emerge from the laboratory with complete self-awareness. However, that story belongs in another edition of *The Best American Science Writing,* many decades from now.)

Some scientists believe that computers based on silicon are just not powerful enough to mimic the human brain. What is needed is to unleash the Ultimate Computer, called the quantum computer, which computes on the ultimate building block of matter: the atom itself. In "Dream Machine," Rivka Galchen explores the promise and problems of building a quantum computer. The stakes are very high: Even the CIA would be interested in quantum computers, since they can break any code devised by any nation. Some have speculated that a quantum computer might instantly change society itself, making possible robots as smart as humans, or allowing us to predict the weather, or unravel how protein molecules fold (the misfoldings of proteins, some suspect, may be implicated in Alzheimer's disease). But alas, quantum computers are notoriously sensitive to any disturbance, rapidly dissipating into a random, incoherent bunch of atoms with the slightest vibration. So the dream machine will have to remain just that—a dream—for the near future.

Moving outward from the body, we then discuss the environment that surrounds us. Moreover, our very survival on this planet rests upon our understanding of how we have degraded it, and whether we can save it.

In "Going to Extremes," Linda Marsa raises a provocative question: Why the wacky weather? Once-in-a-century storms, droughts, and floods seem to be happening all the time. Meteorologists uniformly claim that there is no single smoking gun that explains all these disasters. The weather is incredibly complicated, with many random effects competing with each other. But to some, this wacky weather is consistent with (but does not necessarily prove) global warming. In this point of view, the main by-product of global warming is increasingly violent swings in the weather, with flooding in one region and bone-dry droughts in another. Marsa develops this idea by focusing on the tragic, massive flooding in Australia in 2011.

What the environment needs, some claim, is a white knight to save us from global warming. One leading candidate is nuclear energy, which might allow us to phase out oil and coal plants, which generate greenhouse gases. But 2011 saw the disaster at Fukushima, Japan, which upset all these plans. Images from Fukushima were seared into the public memory. In the wake of the tragedy, Germany and Switzerland have decided to phase out nuclear energy entirely. Even the Japanese government admits that it might take up to forty agonizing years to finally clean up this disaster. In "The Fire Next Time," Jeff Goodell gives us a sobering lesson in what happens when the rosy press releases from the nuclear industry go up in smoke.

One inspiration for the environmental movement has been the majesty and splendor of the earth when we gaze at the iconic pictures of our home world from the space program and see a gorgeous blue sphere floating in the blackness of space. But sadly, the space program that we grew up with is no more. In "The Last Shuttle Launch," P. J. O'Rourke writes a sad farewell to an old and familiar friend: the Space Shuttle. In 2011, the space program was hit with a triple whammy: the cancellation of the Space Shuttle, its replacement (the Constellation program), and the moon and Mars programs. This was a sudden shock to space buffs, who are used to seeing American astronauts greet us from outer space. (Ironically, who would have thought that the fierce rivalry between the United States and Russia would end with the United States hitchhiking on Russian booster rockets?)

Alas, the future of manned space travel rests upon a dirty four-letter word, c-o-s-t. It costs ten thousand dollars to put a pound of anything into near Earth orbit. (That is roughly your weight in gold.) To put you on the moon costs roughly a hundred thousand dollars per pound. And to put you on the Red Planet costs a staggering one million dollars per pound, or roughly your weight in diamonds. Given the enormous cost of space travel, it is no wonder that entrepreneurs are itching to grab a piece of the manned space program, betting that the ingenuity and perseverance of private industry can take us cheaply into space. Perhaps the future belongs not to daring astronauts, but to space tourists spending the weekend in space. That is the vision given by Erik Sofge in "The Early Adopter's Guide to Space Travel." Perhaps private enterprise will boldly go where no government has gone before.

Leaving Earth orbit and soaring into deep space, in "Stellar Oddballs," Charles Petit writes about one of the great revolutions in all of astronomy: the discovery of extra-solar planets. Already, several thousand have been identified by the Kepler satellite. Because an extra-solar planet is found every few days, astronomy textbooks are obsolete as soon as they are published. In fact, we now know that there are more planets than stars in our Milky Way galaxy, so there are more planets in the universe than all the grains of sand on the beach.

Scientists are even finding the imaginary worlds of science fiction (e.g., planets revolving around double and triple star systems, as in *Star Wars*). But since most of these planets are lifeless, Jupiter-sized spheres, the Holy Grail of planetary astronomy is to find a twin of the earth in outer space that can support life. The next time, when we gaze idly at the night sky, we will wonder if anyone is looking back.

But since time immemorial, the ancients who gazed wondrously at the night sky asked an even more profound question: where did it all come from? All great mythologies and religions ask the key question: How was the universe born? Today, physicists believe the key to understanding the origin of the universe lies in the subatomic particles that dominated the Big Bang. The avalanche of subatomic particles found by smashing atoms apart can be arranged into something called the Standard Model, which is remarkably like Mendelev's Periodic Chart of Elements, which enabled chemists to make sense of the thousands of substances found in nature. Similarly, the Standard Model helps physicists make sense of the rich diversity of subatomic particles.

And the key to understanding the Standard Model is the symmetry that allows us to arrange these particles in a sublime order. In "Symmetry: A 'Key to Nature's Secrets'," Nobel Laureate Steven Weinberg shows he is not only a master of physics, but also of great science writing. Artists, poets, and physicists share a common passion: the love of beauty. But while artists and poets are often at a loss to explain this fascination with beauty, physicists are much more precise: beauty is symmetry, which can be described using mathematics. As Weinberg notes, the search for symmetry is not just peripheral to science, but goes to the very heart of physics. Physicists now realize that the symmetry of subatomic particles holds the secret to the universe itself.

So the Standard Model gives us a compelling picture of the symmetry of subatomic particles found so far. Except for one. The last remaining piece of this cosmic jigsaw puzzle is the Higgs particle. "Waiting for the Higgs," by Tim Folger, highlights a ten-billion-dollar machine, the Large Hadron Collider, which was built outside Geneva to find this elusive subatomic particle.

The Standard Model, in turn, represents the ultimate expression of the quantum theory, which governs the subatomic world. The quantum theory is perhaps the most successful theory of all time, proved to within one part in one hundred billion, a breathtaking accuracy unrivaled in all of science. All the gadgets in our labs and living rooms (such as GPS, transistors, lasers, MRIs, and chips) are dependent on the quantum theory. Yet this remarkably successful theory is based on a philosophical foundation of sand. Since quantum mechanics reduces reality to probabilities, it means that bizarre universes straight out of *The Twilight Zone* can never be totally dismissed.

These delicious paradoxes are mentioned in "Moved by Light," by Devin Powell, where he refers to one of the strangest paradoxes in all of science: Schrödinger's cat. According to the quantum theory, if a cat is placed in a sealed box, then, in some sense, the cat can be simultaneously dead and alive. Not surprisingly, Einstein thought this was preposterous, yet electrons exhibit this strange behavior all the time in your laptop, GPS, or iPhone. Experiments have repeatedly shown that electrons can be simultaneously in many states.

But when quantum mechanics is applied to the universe as a whole, then all the fireworks start. In particular, cosmology is forcing physics to confront something straight out of science fiction: parallel universes. In "The Accidental Universe," Alan Lightman writes about one of the strangest upheavals in cosmology: the emergence of the concept of the multiverse, the idea that our universe is not alone. Cosmology is defined as the study of the origin of the universe, but the very word *uni-verse,* meaning "one world," may soon become obsolete, if our universe is just one soap bubble floating in an infinite sea of other bubble-universes.

Last, we enter an area that explores the very limits of science, where it collides with religion and philosophy. Science rests on the bedrock of what is testable, reproducible, and falsifiable. Yet matters of the spirit and the soul are based on faith, miracles, and divine revelation, which, by definition, resist quantification.

There are scientists, however, who have tried to explain the origin of religion from a purely evolutionary point of view, arguing that early humans who adopted a form of religion to keep the tribe together were better able to survive in the wilderness. In this vein, Charles C. Mann, in "The Birth of Religion," tries to trace the origins of religion by using archaeological studies done in southern Turkey, which show that religion is surprisingly ancient, probably predating even the discovery of agriculture.

Still, although science may try to explain the emergence of religion in terms of evolutionary theory, the core philosophical ideas of religion remain outside the grasp of science. I imagine when *The Best American Science Writing 2100* is published, there will still be articles arguing the pros and cons of religion. (To be honest, my own point of view is that it is not possible to logically prove, or disprove, the existence of God. This is a question beyond the province of ordinary science; it is inherently undecidable.)

In Jackson Lears's article, "Same Old New Atheism," he tackles this question from a new, controversial point of view, arguing that atheists within the scientific community, especially since the 9/11 attack on the World Trade Center, are going too far. It is one thing to argue that there is no God, but it goes too far if we criticize entire religions, such as the Muslim religion, and try to elevate science itself into some religion of its own.

Science, of course, is about the rational understanding of the laws of nature. Science is supposed to be logical and mercilessly self-consistent. So how does science confront its precise opposite: madness? In "God Knows Where I Am," Rachel Aviv reminds us of the ethical dilemma when science grapples with the human tragedy of mental illness. Here is the core question: what happens when sick people reject the very medicines that may quell their demons? On one hand, some may argue that individual liberties and freedoms trump the cold, clinical analyses of uncaring psychiatrists. Others argue that it is for people's own good that they are incarcerated against their will and forced to take medicines. (Still others lament the fact that state governments, in order to save on costs, closed down scores of mental hospitals, which left tens of thousands of mentally ill patients without proper care or shelter.) This debate, sadly, is played out in the cold streets of our cities every night.

These are the articles that have fascinated me and educated me, and represent, I feel, some of the best science articles of the past year. And I hope that they will enlighten you as much as they have enlightened me. I hope that they provide some insight into the human condition and this wondrous universe of ours. And, not least, I hope they may inspire young scientists, or young people curious about science, to spark that "Sputnik moment," a new era of innovation and discovery.

The Best American Science Writing 2012

GRETCHEN VOGEL

Mending the Youngest Hearts

FROM *SCIENCE*

Scientists are using stem cells to create blood vessels for damaged hearts. But what surprises them, Gretchen Vogel discovers, is not how effective the technique is, but that the stem cells do not seem to play the role they were once believed to play.

THE NOTION THAT TISSUE ENGINEERS CAN PROVIDE a stock of lab-grown body parts to replace faulty tissues is still, for the most part, a dream. Ready-made hearts, livers, or kidneys that could ease the shortage of donor organs will not be available in the clinic anytime soon. But recent progress suggests the dream is not completely beyond reach. Lab-grown bladders are functioning in dozens of patients in the United States, and doctors in Europe have implanted lab-grown tracheas into several patients.

In Japan, several dozen children and young adults born with severe heart defects are living with tissue-engineered cardiac blood vessels. The first received implants 10 years ago. They go to school, hold full-time jobs, play sports—in short, says Christopher Breuer, one of the implants' developers, they live active, healthy lives. This month, after an arduous approval process, surgeons are testing the blood vessels in the first U.S. patients.

The U.S. trial marks "an important signpost for the whole field," says Joseph Vacanti, a transplant surgeon and tissue engineer at Massachusetts General Hospital in Boston. The implants, which are used to connect a major cardiac vein and the artery that carries blood to the lungs, are made of a synthetic scaffold seeded with cells from the patient's own bone marrow. In the body, the graft develops into a living blood vessel that grows with the patient.

The engineered vessels were developed by a group at Yale University led by Breuer, a pediatric surgeon, and Toshiharu Shinoka, a cardiosurgeon. Although getting approval for the trial from the U.S. Food and Drug Administration (FDA) took more than 4 years and generated more than 3,000 pages of documents, the process paid off, Breuer says: Recent animal studies, arising in part from questions the FDA asked, have turned some of Breuer and Shinoka's assumptions upside down, leading to a better understanding of how the graft works and ideas for how to improve it.

The new experiments suggest that inflammation, long seen as an enemy of transplants and artificial implants alike, seems to play a key role in the transformation of the cell-filled scaffold into a healthy blood vessel. And stem cells, which have been seen as the stars of tissue engineering, play a less significant role than expected. The results are prompting tissue engineers to rethink the role of inflammation and stem cells, says Anita Driessen-Mol, a tissue engineer at Eindhoven University of Technology in the Netherlands. "It's very inspiring work," she says.

Replumbing the Heart

The lab-made blood vessels are meant for children whose severely malformed hearts are unable to supply their bodies with enough oxygen. At birth the children are known as "blue babies" for the skin tint that results. Unlike children with a normal heart, which has two blood-pumping chambers, or ventricles, these children have only one working ventricle. Without a repair, Breuer says, 70% of children with such defects will die before their first birthday.

In the late 1960s, the surgeon Francis Fontan and his colleagues developed a technique to make such hearts more efficient. They rearranged the organ's plumbing to concentrate pumping in the single functioning ventricle. Over the years, surgeons have improved the procedure by adding a length of blood vessel to better connect the heart's inferior vena cava, which collects blood from veins in the lower body, to the pulmonary artery, which leads to the lungs, bypassing the heart. In a few cases, surgeons can build this diversionary vessel from the patient's own tissue. But often there isn't enough tissue available, and surgeons use tubes of synthetic materials such as Gore-Tex.

Such artificial blood vessels have significant drawbacks, Breuer says. The grafts can become calcified, trigger blood clots, and, if cells build up on the inside, can develop stenosis, a dangerous narrowing of the vessel. And because the synthetic graft doesn't grow with the child, surgeons must either delay surgery until the heart has grown larger or implant a graft that is initially too big.

For more than a decade, Breuer and Shinoka have been working to develop lab-grown vessels that act like a patient's own tissue. The team uses a scaffold made of a biodegradable polyester tube, which they incubate briefly with a patient's bone marrow mononuclear cells (BMCs)—a mix of cells including immune cells and blood- and vessel-forming stem cells. Some of the patient's cells stick, seeding the scaffold.

Preclinical experiments in lambs showed that the grafts soon formed a normal-looking blood vessel, with endothelial cells lining the inside and smooth muscle cells surrounding them. As expected, the scaffold degraded within a few months; new collagen fibrils, the connective tissue that helps blood vessels hold their shape, replaced it. And just like a normal blood vessel, the new tissue could grow. Based on those positive results, the clinical trial in Japan was approved and went ahead. Results have been very promising. The only complications were a few cases of stenosis, Shinoka and his colleagues have reported.

Still, researchers weren't sure exactly how the new blood vessel formed. Were the seeded cells growing and differentiating? Or were new cells migrating into the graft? To better understand what happens after implantation—and to prove to the FDA "that each step of the way we were doing what we thought we were," Breuer says, he and his colleagues went back to the lab. Using new, more precise fabrication techniques, they developed a mouse-sized version of their blood vessel scaffold. They seeded it with human BMCs and implanted the vessel in mice. To their surprise, though the vessel remained intact, the human cells disappeared within a week.

To test whether the human cells disappeared because they were attacked by the mouse immune system, the researchers seeded a vessel scaffold with mouse cells genetically matched to the recipient but tagged with green fluorescent protein (GFP). In a paper in *The FASEB Journal* this month, they confirm their earlier results: A week after implantation, almost all of the GFP-tagged cells had disappeared.

The observation "was a huge eye opener for the field," Driessen-Mol says. "We thought that the cells you put in there would still be around for weeks." Instead, it seems, the blood vessel that forms somehow comes from cells in the host's body. Although the seeded cells don't stick around for long, they do provide an advantage to the implant, the researchers reported last year in the *Proceedings of the National Academy of Sciences*. They secrete a protein that attracts monocytes, immune cells that modulate inflammation and can help prompt the formation of new blood vessels. Breuer's team showed that the seeded cells "are essential to initiate the proper kind of inflammation response. Somehow they attract the right kind of initial cells," Driessen-Mol says.

To pinpoint the origin of the cells that build the new blood vessel, the Yale researchers created special chimeric mice. They gave female mice a lethal dose of radiation to destroy the rodents' immune and blood-forming stem cells, then rescued the mice with bone marrow stem cells from males that carried the GFP gene. The donated cells repopulated the animals' bone marrow, and soon the female mice had GFP-tagged male cells in their blood.

The researchers implanted their cell-seeded polyester tube into the chimeric mice and tracked what happened. In the first few weeks after implantation, the researchers found male, GFP-expressing immune cells in the grafts. But by 6 months, the endothelial cells and smooth muscle cells that formed the new stable blood vessel were all female. They also found no sign of stem cell markers in the vessel, suggesting that the cells growing into the graft were differentiated cells from the female host—not her stem cells at all. Further experiments with tagged cells showed that the new cells come from the adjacent blood vessels.

Taken together, Breuer says, the evidence suggests that the graft prompts the adjacent vessels to expand into and over the implanted scaffold in a process similar to normal blood vessel growth. Robert Nerem, a tissue engineer at the Georgia Institute of Technology in Atlanta, says these experiments represent "exactly the kind of study that needs to be done so we understand what's happening mechanistically in these therapeutic approaches." Still, he would like the results confirmed in larger animals.

ENGINEERED REGENERATION

The results from these animal studies overturn the belief, held by many tissue engineers, that rare stem cells in the seeded BMCs would differentiate and grow on the implanted scaffold to form new tissue. The results are consistent with other evidence that BMCs can prompt tissue repair without contributing to the new tissue directly; something similar seems to happen when the cells are injected into diseased hearts. Breuer says that a more precise understanding may help researchers design safer and more effective lab-grown tissues. He and Shinoka are working to develop grafts that wouldn't need seeded cells but instead would contain a combination of signaling molecules to attract the needed response from monocytes.

That could make the implants easier to produce and much less expensive, Vacanti says, increasing the chance that they would be widely adopted. He says the new understanding might also simplify efforts to construct the tissue engineer's ultimate challenge: whole organs, with multiple cell types and a full set of blood vessels. For researchers trying to construct entire hearts, for example, "you could imagine that you wouldn't need to preseed with vascular cells, just with muscle cells," he says.

In the meantime, the new trial will track the six U.S. patients after they receive their implants, following up with regular magnetic resonance imaging scans to watch for signs of stenosis. The team proposed starting small, Breuer says, to signal to the FDA that "we're willing to get started very slowly and carefully." That's the right approach, Vacanti says. "It's terrific that they are going ahead," he says. "Their work is thoughtful, rigorous, and very carefully done." Setting a positive precedent is crucial for the field, he says. "They'll do it properly."

DENISE GRADY

An Immune System Trained to Kill Cancer

FROM THE *NEW YORK TIMES*

A team at the University of Pennsylvania has developed a gene-therapy technique that supercharges the immune system—and has shown that it can put leukemia into remission. Denise Grady investigates whether this new approach might cure cancer.

A YEAR AGO, WHEN CHEMOTHERAPY STOPPED WORKing against his leukemia, William Ludwig signed up to be the first patient treated in a bold experiment at the University of Pennsylvania. Mr. Ludwig, then 65, a retired corrections officer from Bridgeton, N.J., felt his life draining away and thought he had nothing to lose.

Doctors removed a billion of his T-cells—a type of white blood cell that fights viruses and tumors—and gave them new genes that would program the cells to attack his cancer. Then the altered cells were dripped back into Mr. Ludwig's veins.

At first, nothing happened. But after 10 days, hell broke loose in his hospital room. He began shaking with chills. His temperature shot up. His blood pressure shot down. He became so ill that doctors moved him into intensive care and warned that he might die.

His family gathered at the hospital, fearing the worst.

A few weeks later, the fevers were gone. And so was the leukemia.

There was no trace of it anywhere—no leukemic cells in his blood or bone marrow, no more bulging lymph nodes on his CT scan. His doctors calculated that the treatment had killed off two pounds of cancer cells.

A year later, Mr. Ludwig is still in complete remission. Before, there were days when he could barely get out of bed; now, he plays golf and does yard work.

"I have my life back," he said.

Mr. Ludwig's doctors have not claimed that he is cured—it is too soon to tell—nor have they declared victory over leukemia on the basis of this experiment, which involved only three patients. The research, they say, has far to go; the treatment is still experimental, not available outside of studies.

But scientists say the treatment that helped Mr. Ludwig, described recently in *The New England Journal of Medicine* and *Science Translational Medicine*, may signify a turning point in the long struggle to develop effective gene therapies against cancer. And not just for leukemia patients: other cancers may also be vulnerable to this novel approach—which employs a disabled form of H.I.V.-1, the virus that causes AIDS, to carry cancer-fighting genes into the patients' T-cells. In essence, the team is using gene therapy to accomplish something that researchers have hoped to do for decades: train a person's own immune system to kill cancer cells.

Two other patients have undergone the experimental treatment. One had a partial remission: his disease lessened but did not go away completely. Another had a complete remission. All three had had advanced chronic lymphocytic leukemia and had run out of chemo-therapy options. Usually, the only hope for a remission in such cases is a bone-marrow transplant, but these patients were not candidates for it.

Dr. Carl June, who led the research and directs translational medicine in the Abramson Cancer Center at the University of Pennsylvania, said that the results stunned even him and his colleagues, Dr. David L. Porter, Bruce Levine and Michael Kalos. They had hoped to see some benefit but had not dared dream of complete, prolonged remissions. Indeed, when Mr. Ludwig began running fevers, the doctors did not realize at first that it was a sign that his T-cells were engaged in a furious battle with his cancer.

Other experts in the field said the results were a major advance.

"It's great work," said Dr. Walter J. Urba of the Providence Cancer Center and Earle A. Chiles Research Institute in Portland, Ore. He called the patients' recoveries remarkable, exciting and significant. "I feel very positive about this new technology. Conceptually, it's very, very big."

Dr. Urba said he thought the approach would ultimately be used against other types of cancer as well as leukemia and lymphoma. But he cautioned, "For patients today, we're not there yet." And he added the usual scientific caveat: To be considered valid, the results must be repeated in more patients, and by other research teams.

Dr. June called the techniques "a harvest of the information from the molecular biology revolution over the past two decades."

HITTING A GENETIC JACKPOT

To make T-cells search out and destroy cancer, researchers must equip them to do several tasks: recognize the cancer, attack it, multiply and live on inside the patient. A number of research groups have been trying to do this, but the T-cells they engineered could not accomplish all the tasks. As a result, the cells' ability to fight tumors has generally been temporary.

The University of Pennsylvania team seems to have hit all the targets at once. Inside the patients, the T-cells modified by the researchers multiplied to 1,000 to 10,000 times the number infused, wiped out the cancer and then gradually diminished, leaving a population of "memory" cells that can quickly proliferate again if needed.

The researchers said they were not sure which parts of their strategy made it work—special cell-culturing techniques, the use of H.I.V.-1 to carry new genes into the T-cells, or the particular pieces of DNA that they selected to reprogram the T-cells.

The concept of doctoring T-cells genetically was first developed in the 1980s by Dr. Zelig Eshhar at the Weizmann Institute of Science in Rehovot, Israel. It involves adding gene sequences from different sources to enable the T-cells to produce what researchers call chimeric antigen receptors, or CARs—protein complexes that transform the cells into, in Dr. June's words, "serial killers."

Mr. Ludwig's disease, chronic lymphocytic leukemia, is a cancer of B-cells, the part of the immune system that normally produces antibodies to fight infection. All B-cells, whether healthy or leukemic, have on their surfaces a protein called CD19. To treat patients with the disease, the researchers hoped to reprogram their T-cells to find CD19 and attack B-cells carrying it.

But which gene sequences should be used to reprogram the T-cells, from which sources? And how do you insert them?

Various research groups have used different methods. Viruses are often used as carriers (or vectors) to insert DNA into other cells because that kind of genetic sabotage is exactly what viruses normally specialize in doing. To modify their patients' T-cells, Dr. June and his colleagues tried a daring approach: they used a disabled form of H.I.V.-1. They are the first ever to use H.I.V.-1 as the vector in gene therapy for cancer patients (the virus has been used in other diseases).

The AIDS virus is a natural for this kind of treatment, Dr. June said, because it evolved to invade T-cells. The idea of putting any form of the AIDS virus into people sounds a bit frightening, he acknowledged, but the virus used by his team was "gutted" and was no longer harmful. Other researchers had altered and disabled the virus by adding DNA from humans, mice and cows, and from a virus that infects woodchucks and another that infects cows. Each bit was chosen for a particular trait, all pieced together into a vector that Dr. June called a "Rube Goldberg-like solution" and "truly a zoo."

"It incorporates the ability of H.I.V. to infect cells but not to reproduce itself," he said.

To administer the treatment, the researchers collected as many of the patients' T-cells as they could by passing their blood through a machine that removed the cells and returned the other blood components back into the patients' veins. The T-cells were exposed to the vector, which transformed them genetically, and then were frozen. Meanwhile, the patients were given chemotherapy to deplete any remaining T-cells, because the native T-cells might impede the growth of the altered ones. Finally, the T-cells were infused back into the patients.

Then, Dr. June said, "The patient becomes a bioreactor" as the T-cells proliferate, pouring out chemicals called cytokines that cause fever, chills, fatigue and other flulike symptoms.

The treatment wiped out all of the patients' B-cells, both healthy ones and leukemic ones, and will continue to do for as long as the new T-cells persist in the body, which could be forever (and ideally should be, to keep the leukemia at bay). The lack of B-cells means that the patients may be left vulnerable to infection, and they will need periodic infusions of a substance called intravenous immune globulin to protect them.

So far, the lack of B-cells has not caused problems for Mr. Ludwig.

He receives the infusions every few months. He had been receiving them even before the experimental treatment because the leukemia had already knocked out his healthy B-cells.

One thing that is not clear is why Patient 1 and Patient 3 had complete remissions, and Patient 2 did not. The researchers said that when Patient 2 developed chills and fever, he was treated with steroids at another hospital, and the drugs may have halted the T-cells' activity. But they cannot be sure. It may also be that his disease was too severe.

The researchers wrote an entire scientific article about Patient 3, which was published in *The New England Journal of Medicine*. Like the other patients, he also ran fevers and felt ill, but the reaction took longer to set in, and he also developed kidney and liver trouble—a sign of tumor lysis syndrome, a condition that occurs when large numbers of cancer cells die off and dump their contents, which can clog the kidneys. He was given drugs to prevent kidney damage. He had a complete remission.

What the journal article did not mention was that Patient 3 was almost not treated.

Because of his illness and some production problems, the researchers said, they could not produce anywhere near as many altered T-cells for him as they had for the other two patients—only 14 million ("a mouse dose," Dr. Porter said), versus 1 billion for Mr. Ludwig and 580 million for Patient 2. After debate, they decided to treat him anyway.

Patient 3 declined to be interviewed, but he wrote anonymously about his experience for the University of Pennsylvania Web site. When he developed chills and a fever, he said, "I was sure the war was on—I was sure C.L.L. cells were dying."

He wrote that he was a scientist, and that when he was young had dreamed of someday making a discovery that would benefit mankind. But, he concluded, "I never imagined I would be part of the experiment."

When he told Patient 3 that he was in remission, Dr. Porter said, they both had tears in their eyes.

Not Without Danger to Patients

While promising, the new techniques developed by the University of Pennsylvania researchers are not without danger to patients. Engineered T-cells have attacked healthy tissue in patients at other centers. Such a reaction killed a 39-year-old woman with advanced colon cancer in a study at the National Cancer Institute, researchers there reported last year in the journal *Molecular Therapy.*

She developed severe breathing trouble 15 minutes after receiving the T-cells, had to be put on a ventilator and died a few days later. Apparently, a protein target on the cancer cells was also present in her lungs, and the T-cells homed in on it.

Researchers at Memorial Sloan Kettering Cancer Center in New York also reported a death last year in a T-cell trial for leukemia (also published in *Molecular Therapy*). An autopsy found that the patient had apparently died from sepsis, not from the T-cells, but because he died just four days after the infusion, the researchers said they considered the treatment a possible factor.

Dr. June said his team hopes to use T-cells against solid tumors, including some that are very hard to treat, like mesothelioma and ovarian and pancreatic cancer. But possible adverse reactions are a real concern, he said, noting that one of the protein targets on the tumor cells is also found on membranes that line the chest and abdomen. T-cell attacks could cause serious inflammation in those membranes and mimic lupus, a serious autoimmune disease.

Even if the T-cells do not hit innocent targets, there are still risks. Proteins they release could cause a "cytokine storm"— high fevers, swelling, inflammation and dangerously low blood pressure—which can be fatal. Or, if the treatment rapidly kills billions of cancer cells, the debris can damage the kidney and cause other problems.

Even if the new T-cell treatment proves to work, the drug industry will be needed to mass produce it. But Dr. June said the research is being done only at universities, not at drug companies. For the drug industry to take interest, he said, there will have to be overwhelming proof that the treatment is far better than existing ones.

"Then I think they'll jump into it," he said. "My challenge now is to do this in a larger set of patients with randomization, and to show that we have the same effects."

Mr. Ludwig said that when he entered the trial, he had no options left. Indeed, Dr. June said that Mr. Ludwig was "almost dead" from the leukemia, and the effort to treat him was a "Hail Mary."

Mr. Ludwig said: "I don't recall anybody saying there was going to be a remission. I don't think they were dreaming to that extent."

The trial was a Phase 1 study, meaning that its main goal was to find out whether the treatment was safe, and at what dose. Of course, doctors and patients always hope that there will be some benefit, but that was not an official endpoint.

Mr. Ludwig thought that if the trial could buy him six months or a year, it would be worth the gamble. But even if the study did not help him, he felt it would still be worthwhile if he could help the study.

When the fevers hit, he had no idea that might be a good sign. Instead, he assumed the treatment was not working. But a few weeks later, he said that his oncologist, Dr. Alison Loren, told him, "We can't find any cancer in your bone marrow."

Remembering the moment, Mr. Ludwig paused and said, "I got goose bumps just telling you those words."

"I feel wonderful," Mr. Ludwig said during a recent interview. "I walked 18 holes on the golf course this morning."

Before the study, he was weak, suffered repeated bouts of pneumonia and was wasting away. Now, he is full of energy. He has gained 40 pounds. He and his wife bought an R.V., in which they travel with their grandson and nephew. "I feel normal, like I did 10 years before I was diagnosed," Mr. Ludwig said. "This clinical trial saved my life."

Dr. Loren said in an interview, "I hate to say it in that dramatic way, but I do think it saved his life."

Mr. Ludwig said that Dr. Loren told him and his wife something he considered profound.

"She said, 'We don't know how long it's going to last. Enjoy every day,'" Mr. Ludwig recalled.

"That's what we've done ever since."

Kristina Rebelo and Walt Bogdanich

X-Rays and Unshielded Infants

FROM THE *NEW YORK TIMES*

> *At hospitals across the country, untrained or uncertified medical technicians may be overradiating patients—including babies, who are most vulnerable to the effects of radiation. What may be even more shocking, Kristina Rebelo and Walt Bogdanich report, is that crucial legislation on this issue remains stalled in Congress.*

IT WAS WELL AFTER MIDNIGHT WHEN Dr. SALVATORE J. A. Sclafani finally hit the "send" button.

Soon, colleagues would awake to his e-mail, expressing his anguish and shame over the discovery that the tiniest, most vulnerable of all patients—premature babies—had been over-radiated in the department he ran at State University of New York Downstate Medical Center in Brooklyn.

A day earlier, Dr. Sclafani noticed that a newborn had been irradiated from head to toe—with no gonadal shielding—even though only a simple chest X-ray had been ordered.

"I was mortified," he wrote on July 27, 2007. Worse, technologists had given the same baby about 10 of these whole-body X-rays. "Full, unabashed, total irradiation of a neonate," Dr. Sclafani said, adding, "This poor, defenseless baby."

And the problems did not end there. Dr. John Amodio, the hospital's new pediatric radiologist, found that full-body X-rays of premature babies had occurred often, that radiation levels on powerful CT scanners had been set too high for infants, and that babies had been poorly positioned, making it hard for doctors to interpret the images.

The hospital had done the full-body X-rays, known as "babygrams," even though they had been largely discredited because of concerns about the potential harm of radiation on the young. Dr. Sclafani and Dr. Amodio quickly stopped the babygrams and instituted tight controls on how and when radiation was used on babies, according to doctors who work there. But the hospital never reported the problems in the unit to state health officials as required.

A little over a week ago, after *The New York Times* asked about the situation at Downstate, the state health commissioner, Dr. Nirav R. Shah, ordered two offices of the department to investigate.

"Our investigators will pull films, they will examine the medical records and they will interview relevant staff," said Claudia Hutton, the department's director of public affairs. "Our authority to investigate goes basically as far as we need it to go."

The errors at Downstate raise broader questions about the competence, training and oversight of technologists who operate radiological equipment that is becoming increasingly complex and powerful. If technologists could not properly take a simple chest X-ray, how can they be expected to safely operate CT scanners or linear accelerators?

With technologists in many states lightly regulated, or not at all, their own professional group is calling for greater oversight and standards. For 12 years, the American Society of Radiologic Technologists has lobbied Congress to pass a bill that would establish minimum educational and certification requirements, not only for technologists, but also for medical physicists and people in 10 other occupations in medical imaging and radiation therapy.

Yet even with broad bipartisan support, the association said, and the backing of 26 organizations representing more than 500,000 health professionals, Congress has yet to pass what has become known as the CARE bill because, supporters say, it lacks a powerful legislator to champion its cause.

In December 2006, the Senate passed the bill, but Congress adjourned before the House could vote. At the time, the House bill had 135 co-sponsors.

"I would think the public would be outraged that Congress was sitting on what could reduce their radiation exposure," said Dr. Fred Mettler, a radiologist who has investigated and written extensively about radiation accidents.

Individual states decide what standards, if any, radiological workers must meet. Radiation therapists are unregulated in 15 states, imaging technologists in 11 states and medical physicists in 18 states, according to the technologists association. "There are individuals," said Dr. Jerry Reid, executive director of a group that certifies technologists, "who are performing medical imaging and radiation therapy who are not qualified. It is happening right now."

Two months ago, in Michigan—which sets no minimum standards for technologists—the Nuclear Regulatory Commission reported that a large hospital had irradiated the healthy tissue of four cancer patients, three of whom suffered burns, because a technologist repeatedly used the wrong radiological device. "It's amazing to us, knowing the complexity of medical imaging, that there are states that require massage therapists and hairdressers to be licensed, but they have no standards in place for exposing patients to ionizing radiation," said Christine Lung, the technologist association's vice president of government relations.

In New York State, technologists must be licensed and prove that they have passed a professional examination. But there were no continuing education requirements—a provision of the CARE bill—until last year, and regulators usually let hospitals decide whether to discipline technologists. Over the last 10 years, New York health officials say they have not disciplined any of the 20,000 or so licensed technologists for work-related problems.

CHILDREN ARE THE MOST AT RISK

Like many hospitals, SUNY Downstate Medical Center had come to realize that children needed special protection from unnecessary radiation.

Because their cells divide quickly, children are more vulnerable to radiation's effects. And as new ways are found to use radiation in diagnosing and treating injuries and disease, children face an ever-increasing number of radiological procedures. One recent study found that by the age of 18, the average child will have already received more than seven radiological exams.

While the procedures save lives, they are also a source of concern because most scientists believe that the effects of radiation are cumulative—the more radiation a patient receives, the greater the chances of developing cancer. In premature infants, minimizing radiation exposure is especially important because they may require multiple radiological exams for problems like underdeveloped respiratory systems.

In 2007, Dr. Sclafani, the radiology chairman, brought in Dr. Amodio, a highly regarded pediatric radiologist, to oversee diagnostic imaging for children and to evaluate existing practices at Downstate, a large teaching hospital that serves mostly the poor.

Dr. Amodio did not like what he saw. "I have started to compile a list of obvious problems with respect to pediatric images, especially in the neonatal population," he said in a July 26 e-mail to Dr. Sclafani.

A guiding principle for any imaging procedure, regardless of age, is that radiation should be limited—or "coned"—to the area being examined. Yet technologists at Downstate did not always follow that rule. "Improper coning—often entire baby is on radiograph," Dr. Amodio wrote in the first of several bullet points summarizing his findings.

Full-body X-rays of babies are rarely done. "We don't do those anymore," said Dr. Marta Hernanz-Schulman, director of pediatric radiology at Vanderbilt University Medical Center. "If I had an image like that, it would most likely have been a stillborn baby."

Dr. Donald Frush, chief of pediatric radiology at the Duke University School of Medicine, said that failing to properly cone, or collimate, the radiation was rare. "The collimation issue is something that technologists are quite aware of and has been emphasized for decades," Dr. Frush said.

Downstate officials did not say how many inappropriate babygrams were taken. In an interview, Dr. Amodio said he did not know why the technologists had failed to protect the infants, but he surmised that because premature babies are especially fragile, technologists might have been afraid to touch them and "do what was really necessary" to administer proper X-rays. "It is a normal human response," he said.

Asked about the case, Dr. David Keys, a board member of the American College of Medical Physics, said, "It takes less than 15 seconds to collimate a baby," adding: "It could be that the techs at Downstate were too busy. It could be that they were just sloppy or maybe they forgot their training."

In his 2007 e-mail, Dr. Amodio said technologists also failed to shield the gonads, a radiosensitive organ. City and state health codes require shielding for young patients, unless it interferes with a diagnosis, which did not appear to be the case at Downstate.

Other problems, according to Dr. Amodio's e-mail, included using the wrong setting on a radiological device, which caused some premature babies to be "significantly overirradiated."

When Dr. Amodio's findings were reported to the hospital's patient safety committee, its chairman, Dr. Eugene M. Edynak, quickly grasped the seriousness of the situation. "Because of the grave nature of these 'findings,' and the need for immediate correction," Dr. Edynak wrote, "I would like Radiology to present these issues at the next Patient Safety Committee." At the same time, Dr. Edynak noted that radiology management had already begun addressing the problems.

Dr. Sclafani was clearly unsettled by the events. "The past two weeks have been among the most troubled of my career," he wrote at the beginning of an expansive e-mail, sent to members of his department at 1:36 a.m. on July 27.

His greatest disappointment was directed at residents and supervisors for not speaking up about the improper X-rays. "Every film, all dictated, and no one brought this to my attention," Dr. Sclafani said.

In another e-mail, he said he felt "alarmed and ashamed" upon seeing poor imaging techniques. "Excessively irradiating children is something we must have zero tolerance about."

Dr. Sclafani recently took a leave from Downstate to do research. But in an interview last year, he said that his department, with Dr. Amodio's help, had made significant changes, not only in reducing the amount of radiation in CT scans for infants as well as adults, but also in reducing unnecessary scans.

In the past, Dr. Sclafani said, manufacturers had marketed CT scanners based on high-quality images, which often meant more radiation. Referring to Dr. Amodio, he said, "What we learned from John is that sometimes the pretty picture is not what we need."

Dr. Amodio described other department changes, including the use of breast shields for girls and, when possible, substituting an ultrasound, which uses no radiation, for CT scans. In addition, he said, he must personally approve all pediatric CT scans.

Downstate officials, after initially answering questions from The Times last year, have declined to answer any more. In a statement, Ronald Najman, a hospital spokesman, said: "We are working with the New York State Department of Health to re-evaluate the issues raised by our Department of Radiology in 2007, and to ensure that we are in compliance with national and state standards."

Push for Continuing Education

Supporters of the proposed CARE legislation say its continuing-education requirement will keep radiological workers abreast of technological changes. If it passes, "certification and licensure will no longer be a one-time event," said Dr. Geoffrey S. Ibbott, former director of the Radiological Physics Center, a federally financed group that tests radiotherapy equipment for accuracy.

A continuing-education provision might have prevented the over-radiation of 76 patients at a hospital in Missouri—a state that does not regulate its radiological workers. The medical physicist there had selected the wrong calibration tool to set up a highly sophisticated linear accelerator.

Ms. Lung, the vice president of the technologists' group, said that while most people knew that radiation could cause cancer and burn holes in patients, "They don't understand that the last person to see that patient, to position that patient, to make sure that procedure is performed safely is the radiological technologist or radiation therapist."

Jerry Reid, executive director of the American Registry of Radiologic Technologists, a group that certifies technologists, said he was optimistic that the proposed legislation, expected to be introduced in March, would finally pass. Congress, he said, "has shown much more interest in this issue over the last year," in the wake of a series of articles in The Times documenting the harm that can result from radiation mistakes.

But even supporters of the bill say much more needs to be done, including making radiological devices safer and requiring that all mistakes be reported to a single national database.

"We still have to address the culture in many radiology and radiation therapy departments where there is reluctance or outright intimidation that prevents people from reporting errors or potential errors," said Dr. Ibbott. "All of our staff must be empowered to identify errors and situations that could lead to errors without fear of retribution."

The American College of Radiology also recommends that all medical radiology units be professionally accredited, yet many are not.

"In my profession, there is very little room for error and no room for unqualified personnel," said Dr. Steve Goetsch, a medical physicist in California who runs training programs in the field.

Jennifer Couzin-Frankel

Aging Genes

FROM *SCIENCE*

Claims for ways to slow down aging—caloric restriction, or consuming foods that contain resveratrol—have made headlines. But some researchers question whether the sirtuin gene, said to be a key part of this process, is involved in it. Jennifer Couzin-Frankel reports on a scientific battle that involves a falling-out among former colleagues and the fate of a seven-hundred-million-dollar start-up.

SEATTLE, W.A.—THE LUSH, DRIZZLY CAMPUS HERE AT the University of Washington (UW) is 4,000 kilometers from the concrete jungle of the Massachusetts Institute of Technology (MIT) in Cambridge, but Matt Kaeberlein keeps finding himself pulled back to the East Coast institution where his unusual scientific career began. Thirteen years ago, as a graduate student from a little-known western school, he stepped into a highly competitive lab and helped launch a new field in the biology of aging. For the past 7 years, he's been systematically dismantling the building blocks he laid, arguing that some of those early discoveries—and many since then—are wrong.

"It's been an adventure," says Kaeberlein, who turned 40 this year. Wearing jeans and glasses with square metal frames, he comes across as a mix of science nerd and Seattle cool. His fifth-floor office is as neat as they come. "I don't think that I could have ever predicted that things would happen the way they happened," he says.

Kaeberlein's journey began in the lab of MIT professor Leonard Guarente back in 1998, chasing a then-heretical idea in science: that certain genes can prolong life. The work started slowly but captured the attention of nearly everyone in the lab, particularly Guarente, an intense, brilliant biologist. The group churned out a series of influential papers that transformed how scientists and the public think about aging. The idea that life span was a malleable part of biology was no longer science fiction. Discoveries from Guarente's lab linked a set of genes to calorie restriction, which had been known for years to stretch life span in animals. This suggested that drugs to mimic the effects of calorie restriction might not be far behind.

About 10 years ago, Kaeberlein and his cohort of Guarente lab members wrapped up their Ph.D.s and postdocs and scattered. Most went on to start labs of their own at top institutions. From there, the story gets peculiar.

Guarente and some former lab members pushed forward with the new aging genes and vigorously promoted their findings. Others, such as Kaeberlein, experienced nagging doubts that grew with time. Kaeberlein wasn't finding what others were reporting in recent experiments. Outside the Guarente circle, some scientists had similar problems while others reported success.

The result is mass confusion over who's right and who's wrong, and a high-stakes effort to protect reputations, research money, and one of the premier theories in the biology of aging. It's also a story of science gone sour: Several principals have dug in their heels, declined to communicate, and bitterly derided one another. Tensions reached a crescendo in September, when Kaeberlein and colleagues in the United Kingdom published one of their most damning papers yet, finding no effects from a key aging gene in worms and flies.

Almost everyone "is from the same place," the MIT lab run by Guarente, says Stephen Helfand, a fly biologist at Brown University who studies aging and who did not start his career there. What's happening now, he says, is "either Shakespearean or Freudian." Or maybe both.

HOOKED

Kaeberlein was not programmed for a career in science. His father was a postal worker and his mother a homemaker, then an office clerk after his parents divorced. Neither graduated from college.

Kaeberlein spent a few years working for United Parcel Service after finishing high school in Seattle, loading trucks at dawn. He enrolled in community college and moved on to Western Washington University, a state school that overlooks Puget Sound about a half-hour south of the Canadian border. When his wife uprooted to Boston to study marine plant biology at Northeastern University, Kaeberlein tagged along, winding up across the Charles River at MIT. In January of his first year he attended a lecture by Guarente— part of a series by faculty members to attract graduate students to their labs—and was hooked.

"Lenny got up and he talked about how his lab was working on aging using yeast and applying genetics and molecular biology," Kaeberlein recalls. "It wasn't so much the specific story as the idea that you could take what I believe is one of the most complex problems in biology and apply biochemistry and genetics and molecular biology to try and study it. . . . That had just never occurred to me before. It was like, wow, that's really cool." In the spring, he signed on with Guarente.

The lab back then was crackling with electricity. Several years earlier, two graduate students named Brian Kennedy and Nicanor Austriaco Jr. had announced they wanted to study aging. Guarente's focus then was on gene regulation; diving into aging research was high-risk. But life experience had left Kennedy fearless. Less than 2 years before joining the lab, when he was 22 years old, he was hit head-on by a drunk driver traveling the wrong way on a highway without any headlights. That driver was killed. Kennedy spent 6 months in a wheelchair recovering from badly broken legs, a torn diaphragm, and a collapsed lung. He postponed the start of graduate school for a year.

The accident changed him. "You never know what's going to happen in life," Kennedy says now from his perch as president and CEO of the Buck Institute for Research on Aging in Novato, California. "You might as well see what you can achieve and not be afraid of failing." (Austriaco wasn't your average graduate student either; after finishing in the Guarente lab, he went on to become a priest.)

Kennedy and Austriaco focused on yeast, a single-celled fungus. Although Kennedy admits that "I felt like we would have to be lucky to find something about human aging from studying something like yeast," he figured it was worth a shot. In 1995, along with Guarente, Kennedy and Austriaco published the prologue to much of what followed: They found that, when mutated to make it more active, a gene called *SIR4* could extend life span in yeast by as much as 30%. "We were all very excited but also very naïve," Guarente says now. "You find things very new and refreshing, [but] you're not quite certain" what they mean.

Kaeberlein's yeast paper, published in *Genes & Development*, got relatively little attention. But 2 years later, in 2001, MIT postdoc Heidi Tissenbaum and Guarente repeated the work in worms, showing that overexpressing *SIR2* extends life span. Three years after that, Helfand, now at Brown, did the same with *SIR2* in flies. In the insular world of aging biology, sirtuins were suddenly of modest interest.

Two discoveries catapulted them to fame. First, a postdoc in Guarente's lab, Shin-Ichiro Imai, found that *SIR2* could sense the metabolic state of a cell, a topic of long-standing interest in longevity research. Scientists knew that cutting calories significantly altered metabolism, but they didn't understand how that led to longer life. *SIR2* might be a missing domino in the lineup, and that's what Guarente's lab described. Calorie restriction impacts a particular chemical, called NAD, which guides how the cell uses energy. NAD also affects *SIR2*. "So anything that changes the levels of NAD in cells will change the activity of sirtuins," Guarente says. "And one thing that does that is diet." Guarente was proposing the solution to a decades-long puzzle of how cells lived longer on fewer calories. Sirtuins were the answer.

Another leap came in 2003. An ambitious Australian postdoc of Guarente's, David Sinclair, who had recently taken a post at Harvard, reported that resveratrol, a molecule found in red wine, boosted sirtuin activity in a test tube and extended life in yeast. Sinclair presented the resveratrol work at a meeting in the Swiss Alps at the same time it was published in *Nature*. "I've been waiting for this all my life," he told a *New York Times* reporter.

The red wine connection "hit the public consciousness in a way that nothing else in the field has," Kaeberlein says. After all, scientists were not only saying they could find a way to mimic calorie restriction without the untenable diet. They were saying that the treatment of choice might be red wine. How much more appealing could an antiaging prescription get?

NAGGING QUESTIONS

By 2002, Kaeberlein was out of the loop. He had finished his Ph.D. with Guarente and wanted a break from academia. He worked a stint at a start-up biotechnology company, which rented a shuttered video rental store in Waltham, Massachusetts, and filled it with lab benches and equipment purchased at an auction. Soon the venture flopped, Kaeberlein's wife finished her Ph.D., and the two decided to move back west. Kaeberlein landed at UW Seattle, where, it happened, Brian Kennedy had already settled.

One afternoon they met for coffee in the rotunda, an open cafeteria with stained green carpet and white pillars. Kennedy and Kaeberlein reminisced about their years in the Guarente lab. Then they started talking about yeast and *SIR2*. Was *SIR2* really the whole story when it came to yeast aging, they wondered? "We both felt that the field had become very narrowly focused," Kaeberlein says. Publications in top journals nearly all spotlighted *SIR2*. "We know there's got to be other stuff out there."

Finding it seemed an overwhelming task. Yeast have about 6000 genes. Kennedy and Kaeberlein knew they would have to study tens of thousands of yeast cells, each with a different gene deleted, to determine which played important roles in aging. To do that, they'd need to sit over their microscopes, hour after hour, watching daughter cells bud off the mother cell and counting them one by one, until the mother stopped producing them and yeast life, as it's measured, came to a halt.

They forged ahead and used a yeast strain whose genetic background was different from that of the yeast that Kaeberlein had studied in the Guarente lab. Yeast are fickle; not all strains, it turns out, respond the same way to *SIR2* and calorie restriction. One strain in Guarente's lab lived longer when it was calorie restricted but not when it was flooded with *SIR2*; another did exactly the opposite, living longer with extra *SIR2* but not when calories were restricted. Kennedy and Kaeberlein happened to find a strain that responded to both environments, and the young scientists experimented with combining them. What they found suggested that the early work was off the mark: The yeast cells lived far longer when glucose was reduced and *SIR2* was overexpressed than when just one factor was modified. The clincher was that even when *SIR2* was deleted, calorie restriction stretched life span—running counter to the idea that calorie restriction worked by increasing *SIR2* activity.

Piecing this together, Kennedy and Kaeberlein reasoned that *SIR2* overexpression and calorie restriction were acting through different pathways and didn't have much to do with each other, at least in yeast. That would explain the extra-long life when they were combined. "Occam's razor says go with the simplest model that explains the data," Kaeberlein says. They published their findings in 2004 in *PLoS Biology*.

The gauntlet thrown by Guarente's former disciples foreshadowed contretemps to come. For one, there was the problem of comparability. Using genetically different organisms made it difficult to replicate experiments between labs. Even in worms, where there's been a concerted effort to study the same strain originally collected from mushroom compost in Bristol, U.K., in the 1950s, specimen collections evolve independently. Emotions also ran high. Guarente reacted with displeasure to the 2004 paper, Kaeberlein says: "That was when Lenny first got really upset."

Guarente describes being taken aback by the challenge. "Initially, I didn't know what to think," he says, "until I looked really closely and saw the conditions were different." Kaeberlein and Kennedy, he argues, starved their yeast much more aggressively than he had.

Kaeberlein, then a postdoc with Stan Fields, a UW professor who uses yeast as a way to develop new technologies for biological discovery, wouldn't back down. Kennedy saw no need to retreat either. Guarente's argument about different dietary conditions, Kaeberlein says, is a "red herring"; the yeast in the 2004 paper were tested under varying glucose concentrations, including one commonly used by Guarente.

Meanwhile, the sirtuin field was charging ahead. The same year Kennedy and Kaeberlein first raised public doubts about how sirtuins functioned, Sinclair launched a company to capitalize on the molecules and the red wine connection he'd uncovered. He had published a second paper in *Nature* showing that resveratrol also extended life in worms and flies and that it acted through *SIR2*, just like calorie restriction. Sirtris, Sinclair's company, planned to test resveratrol and other sirtuin activators in animals and eventually in people in hopes of preventing disease and ultimately extending life.

But Sinclair would soon have to contend with Kaeberlein and Kennedy. The pair turned back to their pet yeast strain to test resveratrol. They drew a blank. "We went through a year of trying every different concentration, every protocol you can think of, back and forth with David, trying to figure out . . . why we were getting different results," Kaeberlein says. Resveratrol wasn't doing anything—not extending life, not activating *SIR2*. They tested it in the same yeast strains Sinclair was using, to no avail. Sinclair proffered various explanations: The glucose concentrations used to restrict calories or the plastic on the petri dishes might be throwing results off.

Kaeberlein and Kennedy went ahead and published in 2005 in *The Journal of Biological Chemistry*. Their bottom line: Contrary to Sinclair's *Nature* papers, resveratrol did nothing to help yeast cells live longer.

LIFE VERSUS HEALTH

Plenty of scientists are fascinated by yeast, but for the rest of the world the big question has always been what sirtuins do in mammals. The answer appears to be complicated.

Researchers set out to mimic earlier yeast, worm, and fly work in mice, overexpressing the *SIR2* gene—which in mammals is called *SIRT1*—and testing whether it had the same effect. Three groups—one led by Guarente, one by Manuel Serrano of the Spanish National Cancer Research Center, and one by Domenico Accili of Columbia

University—all tried their hand at this. In no study did the mice live longer than usual.

But they weren't your average mice, either; the rodents looked unusually healthy. Guarente's mice, which he described in 2007 in *Aging Cell*, had lower cholesterol and better glucose tolerance and looked a lot like animals deprived of calories. The other groups reported similar results: less type 2 diabetes, healthier metabolic profiles, and healthier livers.

Sirtuin supporters quickly focused on the positives. These mice didn't live longer, but they stayed healthier longer, and wasn't that what mattered to most people? "It's really unclear whether mammalian sirtuins have a role in" extending life, says David Lombard, another Guarente alum now at the University of Michigan, Ann Arbor. "What is undeniable is that sirtuins promote health span" or extended good health. To Lombard, "even if no one showed longevity extension by sirtuins, it doesn't mean they're unimportant."

Skeptics, including Kaeberlein, ask whether life span and health span are really separable. "It's hard for me to imagine a way in which you would slow the progression of multiple age-related diseases without doing something about the molecular damage that is causing the aging process," Kaeberlein says. As it happens, this is a rare point on which Guarente and Kaeberlein agree. His former mentor doesn't believe they can be disentangled, either.

Another doubter is David Harrison, who studies aging in mice at the Jackson Laboratory in Bar Harbor, Maine. Harrison, who is part of a federally funded consortium testing various antiaging compounds in mice, is critical of many mouse studies because they focus on only one strain. "Each mouse strain is a different individual," he says, and individuals often respond differently to the same treatment. The Jackson Lab and its collaborators have so far offered more than 20 compounds to genetically diverse mice to test whether they slow aging.

Resveratrol didn't work, Harrison says. But even skeptics agree that resveratrol and related molecules might help reduce the risk of type 2 diabetes and other metabolic diseases, as well as fatty liver disease—a benefit that's been recorded in mice given the drug. Last month, researchers reported in *Cell Metabolism* that 11 obese men given resveratrol had lower glucose levels, triglycerides, and markers for inflammation in their blood. Sinclair says that resveratrol extends life in mice fed a high-fat diet—but Harrison notes that might be simply because the compound prevented type 2 diabetes.

As for whether sirtuins stretch life in mice and mimic calorie restriction, that puzzle remains unsolved. Unlike people, most mice die of one disease: the cancer lymphoma. And overexpressing *SIRT1* in mice, although it does appear to protect against cardiovascular disease and loss of muscle mass and cognitive decline, doesn't do much to target lymphoma. So the animals still live an average life span, as several experiments, including Guarente's, have shown.

Those studies might not be the last word, however. Shin-Ichiro Imai of Washington University in St. Louis in Missouri is trying to take mouse studies of sirtuins to a new level. Imai, the former Guarente postdoc who helped link *SIR2* to calorie restriction early on, thinks that many past studies are incomplete because *SIRT1* was overexpressed throughout the animals' body. Levels may vary tremendously from one site to another, or from one mouse to another. Imai is trying a different approach: overexpressing *SIRT1* selectively in different tissues. "We do have some very critical results," but they're not yet published, Imai says. "We are 100% convinced that mammalian *SIRT1* plays a role in caloric restriction," and by extension, in aging.

Loose Cannons

Kaeberlein's latest, and arguably most acrimonious, venture into sirtuin land began a few years back, when he ran into a British scientist named David Gems at a conference. Gems studies the biology of aging at University College London and by his own description excels at finding errors. "I've tended to be kind of a loose cannon," Gems confesses, sorting out "pitfalls" that others overlook. In 2005, he heard gossip about worm work Guarente and his postdoc Tissenbaum had published back in 2001, linking *SIR2* to longer worm life. The rumor was that *SIR2*'s potent effect on life span disappeared when genetic differences between control animals and those overexpressing *SIR2* were minimized. Gems hesitated to get involved but in the end chose to investigate.

Based on experiments in his lab, Gems concluded that the rumors were true. So-called outcrossing of the worms—mating them repeatedly with worms from the same strain—smoothes out differences between the genetically modified strain and the control group. Scientists then verify which worms are still overexpressing *SIR2*. The hope is that other genetic variables have been largely erased. Outcrossing the worms up to six times revealed that *SIR2* had no effect on life span.

Gems asked Kaeberlein to try to replicate these results, which a graduate student in Kaeberlein's lab did. Meanwhile, a collaborator of Gems's, Linda Partridge, found the same problem in flies: *SIR2* overexpression, she concluded, didn't have an effect on their life span, either. This contradicted work that Helfand had published in 2009 showing that in genetically identical flies, *SIR2* overexpression extended life. "It's basically a boring little story that says if you do the experiment properly," you arrive at the correct results, Partridge says.

Gems, Partridge, Kaeberlein, and their colleagues published their report in September in *Nature*, the journal in which so much earlier work heralding sirtuins first appeared.

Partridge and Gems took the unusual step of not alerting Guarente, who led the worm work, or Helfand, who led the fly work, before publication of their conflicting findings. Partridge says she challenged some aging work of Helfand's in the past "and got a very dusty response, so I didn't think contact would be helpful." Gems made a similar point. "Normally we would" reach out, he says. But in this case, "based on some of the previous interactions, we thought it would be futile."

Guarente believes otherwise. If he'd heard from his British counterparts, he says, "I think this would have gotten sorted out without dueling papers. That's certainly how we would operate if we found we couldn't reproduce something in someone else's lab." As it happens, a colleague of Guarente's had alerted him to a problem in the worm strain, and he was already looking into it.

Kaeberlein was more conflicted. He encountered Guarente at a meeting in 2010 and updated him, offering to test any worm strains Guarente wanted to supply. Guarente provided specimens, but in Kaeberlein's hands the experiment didn't work. Oddly, some worms with more *SIR2* lived longer than expected, but so did the control animals. "We can't interpret that," Kaeberlein says.

Guarente tested the worms himself. He discovered that unbeknownst to him and Tissenbaum 10 years earlier, the animals carried a second genetic mutation unrelated to sirtuins, and that eliminating it left only about a 10% to 15% extension of life span from *SIR2*, not the 30% reported.

"Nobody's falsifying data; people are just getting different results, and I think there's room for discussion in all of this," says Tissenbaum, now at the University of Massachusetts Medical School in Worcester. She left the sirtuin field several years ago, weary of all the controversy.

"I believe that all the results are likely to be correct," says Felipe Sierra, director of the Division of Aging Biology at the U.S. National Institute on Aging, which has funded much of the academic sirtuin research. "What matters is that there seems to be an effect in some circumstances."

This line of thinking nags at Kaeberlein. Eventually, he says, if you try hard enough, you might be able to extend life span with almost any gene. This doesn't mean that the gene is actually behind aging in a real, living, breathing organism.

And although Kaeberlein agrees that ultimately mammals matter most when it comes to human health, he also believes that the unfolding story line in lower organisms should worry those working with mice and people. "All of this mammalian sirtuin hype is based on the worm and fly work," he says. "Now that that's looking a little questionable, you have to wonder about the rationale for even doing these experiments in the first place."

BILLION-DOLLAR QUESTION

In 2008, GlaxoSmithKline bought Sinclair's company, Sirtris, for $720 million. Sinclair remains a professor at Harvard Medical School and is an adviser to the company. Speaking by phone during a recent trip to Sydney, Australia, Sinclair said he stood by his data. "Rumors of the death of sirtuins and aging are greatly exaggerated," he said. "There are now over 1,000 papers on the subject every year." (A PubMed search of sirtuins found just over 2,000 dating back to 1994.) Sinclair is examining how sirtuin activators impact physiology.

Gems is through with sirtuins. "We never really worked on them anyway; it was just a tidying-up operation," he says. Gems worries about the effect on aging research, even science generally, if sirtuins don't pan out. "If it turns out that this was a giant bubble—how is it possible for so many publications, so much money, it shouldn't have got that far, it shouldn't have happened," he says. He knows of some groups that chose not to publish negative results in the field. "There's a view that seems to be current, that somehow one doesn't engage in quarrels. Sometimes, you have to."

Looking back on those heady days in his lab, Guarente feels "like we were flailing around trying to find some interesting genes." He's still convinced, without a doubt, that he and his mentees did.

"There's an overwhelming case in mammals" that sirtuins are linked to calorie restriction, "and you cannot negate that," he says. He believes, too, that the worm and fly data are correct but that *SIR2* needs to be overexpressed at particular levels to extend life, something that's not easy to do. "I'd believe the positive result," he says. "There's lots of reasons why you can do an experiment and have it not work." About his former student Kaeberlein, he wouldn't say much.

Kennedy, now juggling administrative and research duties in northern California, where he moved from UW last year, agrees that the mammalian work looks promising. But "I have a hard time believing that one protein has been placed on Earth to do positive things in every tissue—we're crossing over into the divine." Still, part of him hopes it will all work out. He finds himself in the odd position, he says, of being the only scientist he knows on the fence about sirtuins.

As for Kaeberlein, his unusual trajectory has taught him as much about how high-stakes science is done as it has about the nitty-gritty of these intriguing molecules. "So much of it has not been hypothesis-driven," he says in frustration. "It's been going in with the idea that these things slow aging." The massive media attention paid to sirtuins has lent them more fame than he believes they deserve. He has no regrets about challenging the dogma, even if he was one of those who pioneered it. "I get that it's embarrassing and people feel bad," he says. But "getting the right answer is more important than people's egos."

Evan Ratliff

Taming the Wild

FROM *NATIONAL GEOGRAPHIC*

For over fifty years, a team of Russian researchers has bred a line of foxes that are as tame as dogs. As Evan Ratliff learns, these experiments have not only shed light on how animals became domesticated; they also provide insights into how humans evolved.

ELLO! HOW ARE YOU DOING?" LYUDMILA TRUT says, reaching down to unlatch the door of a wire cage labeled "Mavrik." We're standing between two long rows of similar crates on a farm just outside the city of Novosibirsk, in southern Siberia, and the 76-year-old biologist's greeting is addressed not to me but to the cage's furry occupant. Although I don't speak Russian, I recognize in her voice the tone of maternal adoration that dog owners adopt when addressing their pets.

Mavrik, the object of Trut's attention, is about the size of a Shetland sheepdog, with chestnut orange fur and a white bib down his front. He plays his designated role in turn: wagging his tail, rolling on his back, panting eagerly in anticipation of attention. In adjacent cages lining either side of the narrow, open-sided shed, dozens of canids do the same, yelping and clamoring in an explosion of fur and unbridled excitement. "As you can see," Trut says above the din, "all of them want human contact." Today, however, Mavrik is the lucky recipient. Trut reaches in and scoops him up, then hands him over to me. Cradled in my arms, gently jawing my hand in his mouth, he's as docile as any lapdog.

Except that Mavrik, as it happens, is not a dog at all. He's a fox. Hidden away on this overgrown property, flanked by birch forests and barred by a rusty metal gate, he and several hundred of his relatives are the only population of domesticated silver foxes in the world. (Most of them are, indeed, silver or dark gray; Mavrik is rare in his chestnut fur.) And by "domesticated" I don't mean captured and tamed, or raised by humans and conditioned by food to tolerate the occasional petting. I mean bred for domestication, as tame as your tabby cat or your Labrador. In fact, says Anna Kukekova, a Cornell researcher who studies the foxes, "they remind me a lot of golden retrievers, who are basically not aware that there are good people, bad people, people that they have met before, and those they haven't." These foxes treat any human as a potential companion, a behavior that is the product of arguably the most extraordinary breeding experiment ever conducted.

It started more than a half century ago, when Trut was still a graduate student. Led by a biologist named Dmitry Belyaev, researchers at the nearby Institute of Cytology and Genetics gathered up 130 foxes from fur farms. They then began breeding them with the goal of re-creating the evolution of wolves into dogs, a transformation that began more than 15,000 years ago.

With each generation of fox kits, Belyaev and his colleagues tested their reactions to human contact, selecting those most approachable to breed for the next generation. By the mid-1960s the experiment was working beyond what he could've imagined. They were producing foxes like Mavrik, not just unafraid of humans but actively seeking to bond with them. His team even repeated the experiment in two other species, mink and rats. "One huge thing that Belyaev showed was the timescale," says Gordon Lark, a University of Utah biologist who studies dog genetics. "If you told me the animal would now come sniff you at the front of the cage, I would say it's what I expect. But that they would become that friendly toward humans that quickly ... wow."

Miraculously, Belyaev had compressed thousands of years of domestication into a few years. But he wasn't just looking to prove he could create friendly foxes. He had a hunch that he could use them to unlock domestication's molecular mysteries. Domesticated animals are known to share a common set of characteristics, a fact documented by Darwin in *The Variation of Animals and Plants Under Domestication*. They tend to be smaller, with floppier ears and curlier tails than their untamed progenitors. Such traits tend to make animals appear appealingly juvenile to humans. Their coats are sometimes spotted—piebald, in scientific terminology—while their wild ancestors' coats are solid. These and other traits, sometimes referred to as the domestication phenotype, exist in varying degrees across a remarkably wide range of species, from dogs, pigs, and cows to some nonmammalians like chickens, and even a few fish.

Belyaev suspected that as the foxes became domesticated, they too might begin to show aspects of a domestication phenotype. He was right again: Selecting which foxes to breed based solely on how well they got along with humans seemed to alter their physical appearance along with their dispositions. After only nine generations, the researchers recorded fox kits born with floppier ears. Piebald patterns appeared on their coats. By this time the foxes were already whining and wagging their tails in response to a human presence, behaviors never seen in wild foxes.

Driving those changes, Belyaev postulated, was a collection of genes that conferred a propensity to tameness—a genotype that the foxes perhaps shared with any species that could be domesticated. Here on the fox farm, Kukekova and Trut are searching for precisely those genes today. Elsewhere, researchers are delving into the DNA of pigs, chickens, horses, and other domesticated species, looking to pinpoint the genetic differences that came to distinguish them from their ancestors. The research, accelerated by the recent advances in rapid genome sequencing, aims to answer a fundamental biological question: "How is it possible to make this huge transformation from wild animals into domestic animals?" says Leif Andersson, a professor of genome biology at Uppsala University, in Sweden. The answer has implications for understanding not just how we domesticated animals, but how we tamed the wild in ourselves as well.

THE EXERCISE OF DOMINION OVER plants and animals is arguably the most consequential event in human history. Along with cultivated agriculture, the ability to raise and manage domesticated fauna—of which wolves were likely the first, but chickens, cattle, and other food species the most important—altered the human diet, paving the way for settlements and eventually nation-states to flourish. By putting humans in close contact with animals, domestication also created vectors for the diseases that shaped society.

Yet the process by which it all happened has remained stubbornly impenetrable. Animal bones and stone carvings can sometimes shed light on the *when* and *where* each species came to live side by side with humans. More difficult to untangle is the *how*. Did a few curious boar creep closer to human populations, feeding off their garbage and with each successive generation becoming a little more a part of our diet? Did humans capture red jungle fowl, the ancestor of the modern chicken, straight from the wild—or did the fowl make the first approach? Out of 148 large mammal species on Earth, why have no more than 15 ever been domesticated? Why have we been able to tame and breed horses for thousands of years, but never their close relative the zebra, despite numerous attempts?

In fact, scientists have even struggled to define domestication precisely. We all know that individual animals can be trained to exist in close contact with humans. A tiger cub fed by hand, imprinting on its captors, may grow up to treat them like family. But that tiger's offspring, at birth, will be just as wild as its ancestors. Domestication, by contrast, is not a quality trained into an individual, but one bred into an entire population through generations of living in proximity to humans. Many if not most of the species' wild instincts have long since been lost. Domestication, in other words, is mostly in the genes.

Yet the borders between domesticated and wild are often fluid. A growing body of evidence shows that historically, domesticated animals likely played a large part in their own taming, habituating themselves to humans before we took an active role in the process. "My working hypothesis," says Greger Larson, an expert on genetics and domestication at Durham University in the United Kingdom, "is that with most of the early animals—dogs first, then pigs, sheep, and goats—there was probably a long period of time of unintentional management by humans." The word domestication "implies something top down, something that humans did intentionally," he says. "But the complex story is so much more interesting."

THE FOX-FARM EXPERIMENT'S ROLE IN unraveling that complexity is all the more remarkable for how it began. The Soviet biology establishment of the mid-20th century, led under Joseph Stalin by the infamous agronomist Trofim Lysenko, outlawed research into Mendelian genetics. But Dmitry Belyaev and his older brother Nikolay, both biologists, were intrigued by the possibilities of the science. "It was his brother's influence that caused him to have this special interest in genetics," Trut says of her mentor. "But these were the times when genetics was considered fake science." When the brothers flouted the prohibition and continued to conduct Mendelian-based studies, Belyaev lost his job as director of the Department of Fur Breeding. Nikolay's fate was more tragic: He was exiled to a labor camp, where he eventually died.

Secretly, Belyaev remained dedicated to genetic science, disguising his work as research in animal physiology. He was particularly consumed with the question of how such an incredible diversity of dogs could have arisen from their wolf ancestors. The answer, he knew, must lie at the molecular level. But even outside the Soviet Union, in the 1950s, the technology to sequence an animal's genome—and thereby try to understand how its genes had changed through history—was an impossible dream. So Belyaev decided to reproduce history himself. The silver fox, a fellow canid and close cousin of dogs that had never been domesticated, seemed the perfect choice.

Lyudmila Trut's first job as a grad student, in 1958, was to travel around to Soviet fur farms and select the calmest foxes she could find, to serve as the base population for Belyaev's experiment. The prohibition on genetic studies had thawed since Stalin's death in 1953, and Belyaev set up shop in Siberia at the newly minted Institute of Cytology and Genetics. Still, he was careful to frame the study only in terms of physiology, leaving out any mention of genes. Trut recalls that when Soviet leader Nikita Khrushchev arrived to inspect the institute, he was overheard to say, "What, are those geneticists still around? Were they not destroyed?" Protected by the careful politics of Belyaev's boss and favorable articles on genetics written by Khrushchev's journalist daughter, the fox-farm experiment quietly began.

By 1964 the fourth generation was already beginning to live up to the researchers' hopes. Trut can still remember the moment when she first saw a fox wag its tail at her approach. Before long, the most tame among them were so doglike that they would leap into researchers' arms and lick their faces. At times the extent of the animals' tameness surprised even the researchers. Once, in the 1970s, a worker took one of the foxes home temporarily as a pet. When Trut visited him, she found the owner taking his fox for walks, unleashed, "just like a dog. I said 'Don't do that, we'll lose it, and it belongs to the institute!'" she recalls. "He said 'just wait,' then he whistled and said, 'Coca!' It came right back."

Simultaneously, more of the foxes began to show signs of the domestication phenotype: floppy ears retained longer in development and characteristic white spots on their coats. "At the beginning of the 1980s, we observed a kind of explosion-like change of the external appearance," says Trut. The research had expanded to include rats in 1972, followed by mink and—for a brief period—river otters. The otters proved difficult to breed and the experiment was eventually abandoned, but the scientists were able to shape the behavior of the other two species in parallel with the foxes.

Just as the genetic tools became available to accomplish Belyaev's end goal of tracing that connection to the animal's DNA, however, the project fell on hard times. With the collapse of the Soviet Union, scientific funds began to dwindle, and the researchers could do little more than keep the fox population alive. When Belyaev died of cancer in 1985, Trut took over the research and fought to keep it funded. But by the beginning of the 21st century, she was in danger of having to shut down the experiment.

Around the same time, Anna Kukekova, a Russian-born postdoc in molecular genetics at Cornell, read about the project's struggles. She had been fascinated with the fox-farm work for years, and now decided to focus her own research on the experiment. With help from Utah's Gordon Lark and a grant from the National Institutes of Health (NIH), she joined Trut's effort to try and finish what Belyaev had started.

NOT ALL THE FOXES ON the farm in Novosibirsk, it turns out, are as friendly as Mavrik. Across the small road from him and his fellow tame foxes is an identical-looking shed full of wire crates, each holding one of what the researchers refer to as the "aggressive foxes." To study the biology of tameness, the scientists needed to create a group of decidedly untame animals. So in a mirror image of the friendly foxes, the kits in the aggressive population are rated according to the hostility of their behavior. Only the most aggressive are bred for the next generation. Here are the evil twins of the tail-wagging Mavrik, straight out of a B-grade horror film: hissing, baring their teeth, snapping at the front of their cages when any human approaches.

"I'd like to draw your attention to this fox," says Trut, pointing to one snarling creature nearby. "You can see how aggressive she is. She was born to an aggressive mother but brought up by a tame mother." The switch, the result of the aggressive mother being unable to feed its kit, serendipitously proved a point: The foxes' response to humans is more nature than it is nurture. "Here," she says, "it's the genetics that change."

Identifying the precise genetic footprint involved in tameness, however, is proving extremely tricky science. First the researchers need to find the genes responsible for creating friendly and aggressive behaviors. Such general behavior traits, however, are actually amalgamations of more specific ones—fear, boldness, passivity, curiosity—that must be teased apart, measured, and traced to individual genes or sets of genes working in combination. Once those genes are identified, the researchers can test whether the ones influencing behavior are also behind the floppy ears and piebald coats and other features that characterize domesticated species. One theory among the scientists in Novosibirsk is that the genes guiding the animals' behavior do so by altering chemicals in their brains. Changes to those neurochemicals, in turn, have "downstream" impacts on the animals' physical appearance.

For now, though, Kukekova is focused on the first step: linking tame behavior to genes. Toward the end of every summer, she travels from Cornell to Novosibirsk to evaluate the year's newborn kits. Each researcher's interaction with a kit is standardized and video-taped: opening a cage, reaching a hand in, touching the fox. Later, Kukekova reviews the tapes, using objective measures to quantify the foxes' postures, vocalizations, and other behaviors. Those data are layered on top of a pedigree—records that keep track of tame, aggressive, and "crossed" foxes (those with parents from each group).

The joint American-Russian research team then extracts DNA from blood samples of each fox in the study and scans for stark differences in the genomes of those that scored as aggressive or tame in the behavioral measures. In a paper in press in *Behavior Genetics*, the group reports finding two regions that are widely divergent in the two behavioral types and might thus harbor key domestication genes. Increasingly, it appears that domestication is driven not by a single gene but a suite of genetic changes. "Domestication," the paper concludes, "appears to be a very complex phenotype."

As it happens, 2,800 miles to the west in Leipzig, Germany, another laboratory is at the exact same juncture in understanding domestication genes in rats. Frank Albert, a researcher at the Max Planck Institute for Evolutionary Anthropology, obtained 30 descendants of Belyaev's rats (15 tame, 15 aggressive) in two wooden boxes from Siberia in 2004. "What we found were regions of the genome that influence tameness and aggression," says Albert. "But we don't know which genes cause these signals." Like Kukekova's group, he says, "we are in the process of whittling down the number."

Once either group is able to pinpoint one or more of the specific genetic pathways involved, they or other researchers can look for parallel genes in other domesticated species. "In a perfect situation, we'd like to define specific genes involved in tame and aggressive behaviors," says Kukekova. "Even when we find those, we will not know if they are the genes for domestication until we compare them in other animals."

Ultimately, the biggest payoff of the research may come from finding similar genes in the most thoroughly domesticated species of all: human beings. "Understanding what has changed in these animals is going to be incredibly informative," says Elaine Ostrander, of the National Human Genome Research Institute at NIH. "Everyone is waiting with great excitement for what they come out with."

NOT ALL DOMESTICATION RESEARCHERS BELIEVE that Belyaev's silver foxes will unlock the secrets of domestication. Uppsala University's Leif Andersson, who studies the genetics of farm animals—and who lauds Belyaev and his fellow researchers' contribution to the field—believes that the relationship between tameness and the domestication phenotype may prove to be less direct than the fox study implies. "You select on one trait and you see changes in other traits," Andersson says, but "there has never been proven a causal relationship."

To understand how Andersson's view differs from that of the researchers in Novosibirsk, it's helpful to try and imagine how the two theories might have played out historically. Both would agree that the animals most likely to be domesticated were those predisposed to human contact. Some mutation, or collection of mutations, in their DNA caused them to be less afraid of humans, and thus willing to live closer to them. Perhaps they fed off human refuse or benefited from inadvertent shelter from predators. At some point humans saw some benefit in return from these animal neighbors and began helping that process along, actively selecting for the most amenable ones and breeding them. "At the beginning of the domestication process, only natural selection was at work," as Trut puts it. "Down the road, this natural selection was replaced with artificial selection."

Where Andersson differs is in what happened next. If Belyaev and

Trut are correct, the self-selection and then human selection of less fearful animals carried with it other components of the domestication phenotype, such as curly tails and smaller bodies. In Andersson's view, that theory understates the role humans played in selecting those other traits. Sure, curiosity and lack of fear may have started the process, but once animals were under human control, they were also protected from wild predators. Random mutations for physical traits that might quickly have been weeded out in the wild, like white spots on a dark coat, were allowed to persist. Then they flourished, in part because, well, people liked them. "It wasn't that the animals behaved differently," as Andersson says, "it's just that they were cute."

In 2009 Andersson bolstered his theory by comparing mutations in coat-color genes between several varieties of domesticated and wild pigs. The results, he reported, "demonstrate that early farmers intentionally selected pigs with novel coat coloring. Their motivations could have been as simple as a preference for the exotic or selection for reduced camouflage."

In his own hunt for domestication genes, Andersson is taking a close look at the most populous domesticated animal on Earth: the chicken. Their ancestors, red jungle fowl, roamed freely in the jungles of India, Nepal, and other parts of South and Southeast Asia. Somewhere around 8,000 years ago, humans started breeding them for food. Last year Andersson and his colleagues compared the full genomes of domesticated chickens with those of zoo-based populations of red jungle fowl. They identified a mutation, in a gene known as *TSHR,* that was found only in domestic populations. The implication is that *TSHR* thereby played some role in domestication, and now the team is working to determine exactly what the *TSHR* mutation controls. Andersson hypothesizes that it could play a role in the birds' reproductive cycles, allowing chickens to breed more frequently in captivity than red jungle fowl do in the wild—a trait early farmers would have been eager to perpetuate. The same difference exists between wolves, which reproduce once a year and in the same season, and dogs, which can breed multiple times a year, in any season.

IF ANDERSSON'S THEORY IS CORRECT, it may turn out to have intriguing implications for our own species. Harvard biologist Richard Wrangham has theorized that we, too, went through a domestication process that altered our biology. "The question of what is the difference between the domestic pig and a wild boar, or the distinction between a broiler chicken and a wild jungle fowl," Andersson told me, "is very similar to the question of what is the difference between a human and a chimpanzee."

Human beings are not simply domesticated chimpanzees, but understanding the genetics of domestication in chickens, dogs, and pigs may still tell us a surprising amount about the sources of our own social behavior. That's one reason the fox-farm research being conducted by Kukekova is underwritten by the NIH. "There are over 14,000 genes expressed in the brain, and not many are understood," she points out. Ferreting out which of those genes are related to social behavior is a tricky business; obviously one cannot perform breeding experiments on humans, and studies purporting to find innate differences in behavior among people or populations are at the very least problematic.

But delving into the DNA of our closest companions can deliver some tantalizing insights. In 2009 UCLA biologist Robert Wayne led a study comparing the wolf and dog genomes. The finding that made headlines was that dogs originated from gray wolves not in East Asia, as other researchers had argued, but in the Middle East. Less noticed by the press was a brief aside in which Wayne and his colleagues identified a particular short DNA sequence, located near a gene called *WBSCR17*, that was very different in the two species. That region of the genome, they suggested, could be a potential target for "genes that are important in the early domestication of dogs." In humans, the researchers went on to note, *WBSCR17* is at least partly responsible for a rare genetic disorder called Williams-Beuren syndrome. Williams-Beuren is characterized by elfin features, a shortened nose bridge, and "exceptional gregariousness"—its sufferers are often overly friendly and trusting of strangers.

After the paper was published, Wayne says, "the number one email we got was from parents of children suffering from Williams-Beuren. They said, Actually our children remind us of dogs in terms of their ability to read behavior and their lack of social barriers in their behavior." The elfin traits also seemed to correspond to aspects of the domestication phenotype. Wayne cautions against making one-to-one parallels between domestication genes and something as genetically complex as Williams-Beuren. The researchers are "intrigued," he says, and hoping to explore the connection further.

In 2003 a young researcher at Duke University named Brian Hare traveled out to Novosibirsk. Hare is known for his work cataloging the unique behaviors of dogs and wolves, showing the ways in which dogs have evolved to follow human cues like pointing and eye movements. When he conducted similar tests on fox kits in Siberia, he found that they did just as well as puppies of the same age. The results, while preliminary, suggest that selecting against fear and aggression—what Hare calls "emotional reactivity"—has created foxes that are not just tame but that also have the doglike ability to engage with humans using their social cues.

"They didn't select for a smarter fox but for a nice fox," says Hare. "But they ended up getting a smart fox." This research also has implications for the origins of human social behavior. "Are we domesticated in the sense of dogs? No. But I am comfortable saying that the first thing that has to happen to get a human from an apelike ancestor is a substantial increase in tolerance toward one another. There had to be a change in our social system."

Hare's research came to mind on my last afternoon in Novosibirsk, as Kukekova, my translator Luda Mekertycheva, and I played with Mavrik in a pen behind the fox farm's research house. We watched him chase a ball and wrestle with another fox, then run back so we could grab him up and let him lick our faces. But we all had flights to catch, and after an hour, Kukekova carried him back toward the sheds. Mavrik seemed to sense that he was headed back to his cage and whined with increasing agitation. Here was an animal biologically conditioned for human attention, as much as any dog is. Now that we'd provided it, I suddenly felt guilty for taking it away.

The fox-farm experiment is, of course, just that: a scientific experiment. For decades the project has been forced to manage their population by selling off to real fur farms those foxes not friendly or aggressive enough to be research candidates. For the scientists, deciding which ones stay and which ones go is a harrowing process; Trut says she has long since passed on the job to others and stays away from the farm during selection time. "It is very difficult emotionally," she told me.

In recent years the institute has been working to obtain permits to sell the surplus tame foxes as pets, both domestically and in other countries. It would be a way not just to find a better home for the unwanted foxes, they suggest, but also to raise money for the research to continue. "The situation today is we are just doing our best to preserve our population," Trut says. "We do some genetic work with our partners in America. But this experiment has many more questions to resolve."

As for Mavrik, Luda Mekertycheva was so enthralled by the chestnut-colored fox and another playmate that she decided to adopt them. They arrived at her dacha outside of Moscow a few months later, and not long after, she emailed me an update. "Mavrik and Peter jump on my back when I kneel to give them food, sit when I pet them, and take vitamins from my hand," she wrote. "I love them a lot."

David Dobbs

Beautiful Brains

FROM *NATIONAL GEOGRAPHIC*

Teenagers are famously reckless, moody, unstable. Recent brain research has attributed their behavior to the fact that their brains are still developing. Prompted by his own son's experiences, David Dobbs explores the current thinking about this kind of behavior: that it may reflect an adolescent's healthy attempts to adapt to becoming independent.

ONE FINE MAY MORNING NOT LONG AGO MY OLDEST son, 17 at the time, phoned to tell me that he had just spent a couple hours at the state police barracks. Apparently he had been driving "a little fast." What, I asked, was "a little fast"? Turns out this product of my genes and loving care, the boy-man I had swaddled, coddled, cooed at, and then pushed and pulled to the brink of manhood, had been flying down the highway at 113 miles an hour.

"That's more than a little fast," I said.

He agreed. In fact, he sounded somber and contrite. He did not object when I told him he'd have to pay the fines and probably for a lawyer. He did not argue when I pointed out that if anything happens at that speed—a dog in the road, a blown tire, a sneeze—he dies. He was in fact almost irritatingly reasonable. He even proffered that the cop did the right thing in stopping him, for, as he put it, "We can't all go around doing 113."

He did, however, object to one thing. He didn't like it that one of the several citations he received was for reckless driving.

"Well," I huffed, sensing an opportunity to finally yell at him, "what would you call it?"

"It's just not accurate," he said calmly. " 'Reckless' sounds like you're not paying attention. But I was. I made a deliberate point of doing this on an empty stretch of dry interstate, in broad daylight, with good sight lines and no traffic. I mean, I wasn't just gunning the thing. I was driving.

"I guess that's what I want you to know. If it makes you feel any better, I was really focused."

Actually, it did make me feel better. That bothered me, for I didn't understand why. Now I do.

My son's high-speed adventure raised the question long asked by people who have pondered the class of humans we call teenagers: What on Earth was he doing? Parents often phrase this question more colorfully. Scientists put it more coolly. They ask, What can explain this behavior? But even that is just another way of wondering, What is wrong with these kids? Why do they act this way? The question passes judgment even as it inquires.

Through the ages, most answers have cited dark forces that

uniquely affect the teen. Aristotle concluded more than 2,300 years ago that "the young are heated by Nature as drunken men by wine." A shepherd in William Shakespeare's *The Winter's Tale* wishes "there were no age between ten and three-and-twenty, or that youth would sleep out the rest; for there is nothing in the between but getting wenches with child, wronging the ancientry, stealing, fighting." His lament colors most modern scientific inquiries as well. G. Stanley Hall, who formalized adolescent studies with his 1904 *Adolescence: Its Psychology and Its Relations to Physiology, Anthropology, Sociology, Sex, Crime, Religion and Education,* believed this period of "storm and stress" replicated earlier, less civilized stages of human development. Freud saw adolescence as an expression of torturous psychosexual conflict; Erik Erikson, as the most tumultuous of life's several identity crises. Adolescence: always a problem.

Such thinking carried into the late 20th century, when researchers developed brain-imaging technology that enabled them to see the teen brain in enough detail to track both its physical development and its patterns of activity. These imaging tools offered a new way to ask the same question—What's wrong with these kids?—and revealed an answer that surprised almost everyone. Our brains, it turned out, take much longer to develop than we had thought. This revelation suggested both a simplistic, unflattering explanation for teens' maddening behavior—and a more complex, affirmative explanation as well.

THE FIRST FULL SERIES OF scans of the developing adolescent brain—a National Institutes of Health (NIH) project that studied over a hundred young people as they grew up during the 1990s—showed that our brains undergo a massive reorganization between our 12th and 25th years. The brain doesn't actually grow very much during this period. It has already reached 90 percent of its full size by the time a person is six, and a thickening skull accounts for most head growth afterward. But as we move through adolescence, the brain undergoes extensive remodeling, resembling a network and wiring upgrade.

For starters, the brain's axons—the long nerve fibers that neurons use to send signals to other neurons—become gradually more insulated with a fatty substance called myelin (the brain's white matter), eventually boosting the axons' transmission speed up to a hundred times. Meanwhile, dendrites, the branchlike extensions that neurons use to receive signals from nearby axons, grow twiggier, and the most heavily used synapses—the little chemical junctures across which axons and dendrites pass notes—grow richer and stronger. At the same time, synapses that see little use begin to wither. This synaptic pruning, as it is called, causes the brain's cortex—the outer layer of gray matter where we do much of our conscious and complicated thinking—to become thinner but more efficient. Taken together, these changes make the entire brain a much faster and more sophisticated organ.

This process of maturation, once thought to be largely finished by elementary school, continues throughout adolescence. Imaging work done since the 1990s shows that these physical changes move in a slow wave from the brain's rear to its front, from areas close to the brain stem that look after older and more behaviorally basic functions, such as vision, movement, and fundamental processing, to the evolutionarily newer and more complicated thinking areas up front. The corpus callosum, which connects the brain's left and right hemispheres and carries traffic essential to many advanced brain functions, steadily thickens. Stronger links also develop between the hippocampus, a sort of memory directory, and frontal areas that set goals and weigh different agendas; as a result, we get better at integrating memory and experience into our decisions. At the same time, the frontal areas develop greater speed and richer connections, allowing us to generate and weigh far more variables and agendas than before.

When this development proceeds normally, we get better at balancing impulse, desire, goals, self-interest, rules, ethics, and even altruism, generating behavior that is more complex and, sometimes at least, more sensible. But at times, and especially at first, the brain does this work clumsily. It's hard to get all those new cogs to mesh.

Beatriz Luna, a University of Pittsburgh professor of psychiatry who uses neuroimaging to study the teen brain, used a simple test that illustrates this learning curve. Luna scanned the brains of children, teens, and twentysomethings while they performed an antisaccade task, a sort of eyes-only video game where you have to stop yourself from looking at a suddenly appearing light. You view a screen on which the red crosshairs at the center occasionally disappear just as a light flickers elsewhere on the screen. Your instructions are to not look at the light and instead to look in the opposite direction. A sensor detects any eye movement. It's a tough assignment, since flickering lights naturally draw our attention. To succeed, you must override both a normal impulse to attend to new information and curiosity about something forbidden. Brain geeks call this response inhibition.

Ten-year-olds stink at it, failing about 45 percent of the time. Teens do much better. In fact, by age 15 they can score as well as adults if they're motivated, resisting temptation about 70 to 80 percent of the time. What Luna found most interesting, however, was not those scores. It was the brain scans she took while people took the test. Compared with adults, teens tended to make less use of brain regions that monitor performance, spot errors, plan, and stay focused—areas the adults seemed to bring online automatically. This let the adults use a variety of brain resources and better resist temptation, while the teens used those areas less often and more readily gave in to the impulse to look at the flickering light—just as they're more likely to look away from the road to read a text message.

If offered an extra reward, however, teens showed they could push those executive regions to work harder, improving their scores. And by age 20, their brains respond to this task much as the adults' do.

Luna suspects the improvement comes as richer networks and faster connections make the executive region more effective.

These studies help explain why teens behave with such vexing inconsistency: beguiling at breakfast, disgusting at dinner; masterful on Monday, sleepwalking on Saturday. Along with lacking experience generally, they're still learning to use their brain's new networks. Stress, fatigue, or challenges can cause a misfire. Abigail Baird, a Vassar psychologist who studies teens, calls this neural gawkiness—an equivalent to the physical awkwardness teens sometimes display while mastering their growing bodies.

The slow and uneven developmental arc revealed by these imaging studies offers an alluringly pithy explanation for why teens may do stupid things like drive at 113 miles an hour, aggrieve their ancientry, and get people (or get gotten) with child: They act that way because their brains aren't done! You can see it right there in the scans!

This view, as titles from the explosion of scientific papers and popular articles about the "teen brain" put it, presents adolescents as "works in progress" whose "immature brains" lead some to question whether they are in a state "akin to mental retardation."

The story you're reading right now, however, tells a different scientific tale about the teen brain. Over the past five years or so, even as the work-in-progress story spread into our culture, the discipline of adolescent brain studies learned to do some more-complex thinking of its own. A few researchers began to view recent brain and genetic findings in a brighter, more flattering light, one distinctly colored by evolutionary theory. The resulting account of the adolescent brain—call it the adaptive-adolescent story—casts the teen less as a rough draft than as an exquisitely sensitive, highly adaptable creature wired almost perfectly for the job of moving from the safety of home into the complicated world outside.

This view will likely sit better with teens. More important, it sits better with biology's most fundamental principle, that of natural selection. Selection is hell on dysfunctional traits. If adolescence is essentially a collection of them—angst, idiocy, and haste; impulsive-ness, selfishness, and reckless bumbling—then how did those traits survive selection? They couldn't—not if they were the period's most fundamental or consequential features.

The answer is that those troublesome traits don't really character-ize adolescence; they're just what we notice most because they annoy us or put our children in danger. As B. J. Casey, a neuroscientist at Weill Cornell Medical College who has spent nearly a decade applying brain and genetic studies to our understanding of adolescence, puts it, "We're so used to seeing adolescence as a problem. But the more we learn about what really makes this period unique, the more adoles-cence starts to seem like a highly functional, even adaptive period. It's exactly what you'd need to do the things you have to do then."

To SEE PAST THE DISTRACTING, dopey teenager and glimpse the adaptive adolescent within, we should look not at specific, some-times startling, behaviors, such as skateboarding down stairways or dating fast company, but at the broader traits that underlie those acts.

Let's start with the teen's love of the thrill. We all like new and exciting things, but we never value them more highly than we do during adolescence. Here we hit a high in what behavioral scientists call sensation seeking: the hunt for the neural buzz, the jolt of the unusual or unexpected.

Seeking sensation isn't necessarily impulsive. You might plan a sensation-seeking experience—a skydive or a fast drive—quite de-liberately, as my son did. Impulsivity generally drops throughout life, starting at about age 10, but this love of the thrill peaks at around age 15. And although sensation seeking can lead to danger-ous behaviors, it can also generate positive ones: The urge to meet more people, for instance, can create a wider circle of friends, which generally makes us healthier, happier, safer, and more successful.

This upside probably explains why an openness to the new,

though it can sometimes kill the cat, remains a highlight of adolescent development. A love of novelty leads directly to useful experience. More broadly, the hunt for sensation provides the inspiration needed to "get you out of the house" and into new terrain, as Jay Giedd, a pioneering researcher in teen brain development at NIH, puts it.

Also peaking during adolescence (and perhaps aggrieving the ancientry the most) is risk-taking. We court risk more avidly as teens than at any other time. This shows reliably in the lab, where teens take more chances in controlled experiments involving everything from card games to simulated driving. And it shows in real life, where the period from roughly 15 to 25 brings peaks in all sorts of risky ventures and ugly outcomes. This age group dies of accidents of almost every sort (other than work accidents) at high rates. Most long-term drug or alcohol abuse starts during adolescence, and even people who later drink responsibly often drink too much as teens. Especially in cultures where teenage driving is common, this takes a gory toll: In the U.S., one in three teen deaths is from car crashes, many involving alcohol.

Are these kids just being stupid? That's the conventional explanation: They're not thinking, or by the work-in-progress model, their puny developing brains fail them.

Yet these explanations don't hold up. As Laurence Steinberg, a developmental psychologist specializing in adolescence at Temple University, points out, even 14- to 17-year-olds—the biggest risk takers—use the same basic cognitive strategies that adults do, and they usually reason their way through problems just as well as adults. Contrary to popular belief, they also fully recognize they're mortal. And, like adults, says Steinberg, "teens actually overestimate risk."

So if teens think as well as adults do and recognize risk just as well, why do they take more chances? Here, as elsewhere, the problem lies less in what teens lack compared with adults than in what they have more of. Teens take more risks not because they don't understand the dangers but because they weigh risk versus reward differently: In situations where risk can get them something they want, they value the reward more heavily than adults do.

A video game Steinberg uses draws this out nicely. In the game, you try to drive across town in as little time as possible. Along the way you encounter several traffic lights. As in real life, the traffic lights sometimes turn from green to yellow as you approach them, forcing a quick go-or-stop decision. You save time—and score more points—if you drive through before the light turns red. But if you try to drive through the red and don't beat it, you lose even more time than you would have if you had stopped for it. Thus the game rewards you for taking a certain amount of risk but punishes you for taking too much.

When teens drive the course alone, in what Steinberg calls the emotionally "cool" situation of an empty room, they take risks at about the same rates that adults do. Add stakes that the teen cares about, however, and the situation changes. In this case Steinberg added friends: When he brought a teen's friends into the room to watch, the teen would take twice as many risks, trying to gun it through lights he'd stopped for before. The adults, meanwhile, drove no differently with a friend watching.

To Steinberg, this shows clearly that risk-taking rises not from puny thinking but from a higher regard for reward.

"They didn't take more chances because they suddenly downgraded the risk," says Steinberg. "They did so because they gave more weight to the payoff."

Researchers such as Steinberg and Casey believe this risk-friendly weighing of cost versus reward has been selected for because, over the course of human evolution, the willingness to take risks during this period of life has granted an adaptive edge. Succeeding often requires moving out of the home and into less secure situations. "The more you seek novelty and take risks," says Baird, "the better you do." This responsiveness to reward thus works like the desire for new sensation: It gets you out of the house and into new turf.

As Steinberg's driving game suggests, teens respond strongly to social rewards. Physiology and evolutionary theory alike offer explanations for this tendency. Physiologically, adolescence brings a peak in the brain's sensitivity to dopamine, a neurotransmitter that appears to prime and fire reward circuits and aids in learning patterns and making decisions. This helps explain the teen's quickness of learning and extraordinary receptivity to reward—and his keen, sometimes melodramatic reaction to success as well as defeat.

The teen brain is similarly attuned to oxytocin, another neural hormone, which (among other things) makes social connections in particular more rewarding. The neural networks and dynamics associated with general reward and social interactions overlap heavily. Engage one, and you often engage the other. Engage them during adolescence, and you light a fire.

This helps explain another trait that marks adolescence: Teens prefer the company of those their own age more than ever before or after. At one level, this passion for same-age peers merely expresses in the social realm the teen's general attraction to novelty: Teens offer teens far more novelty than familiar old family does.

Yet teens gravitate toward peers for another, more powerful reason: to invest in the future rather than the past. We enter a world made by our parents. But we will live most of our lives, and prosper (or not) in a world run and remade by our peers. Knowing, understanding, and building relationships with them bears critically on success. Socially savvy rats or monkeys, for instance, generally get the best nesting areas or territories, the most food and water, more allies, and more sex with better and fitter mates. And no species is more intricately and deeply social than humans are.

This supremely human characteristic makes peer relations not a sideshow but the main show. Some brain-scan studies, in fact, suggest that our brains react to peer exclusion much as they respond to threats to physical health or food supply. At a neural level, in other words, we perceive social rejection as a threat to existence. Knowing this might make it easier to abide the hysteria of a 13-year-old deceived by a friend or the gloom of a 15-year-old not invited to a party. These people! we lament. They react to social ups and downs as if their fates depended upon them! They're right. They do.

EXCITEMENT, NOVELTY, RISK, THE COMPANY of peers. These traits may seem to add up to nothing more than doing foolish new stuff with friends. Look deeper, however, and you see that these traits that define adolescence make us more adaptive, both as individuals and as a species. That's doubtless why these traits, broadly defined, seem to show themselves in virtually all human cultures, modern or tribal. They may concentrate and express themselves more starkly in modern Western cultures, in which teens spend so much time with each other. But anthropologists have found that virtually all the world's cultures recognize adolescence as a distinct period in which adolescents prefer novelty, excitement, and peers. This near-universal recognition sinks the notion that it's a cultural construct.

Culture clearly shapes adolescence. It influences its expression and possibly its length. It can magnify its manifestations. Yet culture does not create adolescence. The period's uniqueness rises from genes and developmental processes that have been selected for over thousands of generations because they play an amplified role during this key transitional period: producing a creature optimally primed to leave a safe home and move into unfamiliar territory.

The move outward from home is the most difficult thing that humans do, as well as the most critical—not just for individuals but for a species that has shown an unmatched ability to master challenging new environments. In scientific terms, teenagers can be a pain in the ass. But they are quite possibly the most fully, crucially adaptive human beings around. Without them, humanity might not have so readily spread across the globe.

THIS ADAPTIVE-ADOLESCENCE VIEW, HOWEVER ACCURATE, can be tricky to come to terms with—the more so for parents dealing with teens in their most trying, contrary, or flat-out scary moments. It's reassuring to recast worrisome aspects as signs of an organism learning how to negotiate its surroundings. But natural selection swings a sharp edge, and the teen's sloppier moments can bring unbearable consequences. We may not run the risk of being killed in ritualistic battles or being eaten by leopards, but drugs, drinking, driving, and crime take a mighty toll. My son lives, and thrives, sans car, at college. Some of his high school friends, however, died during their driving experiments. Our children wield their adaptive plasticity amid small but horrific risks.

We parents, of course, often stumble too, as we try to walk the blurry line between helping and hindering our kids as they adapt to adulthood. The United States spends about a billion dollars a year on programs to counsel adolescents on violence, gangs, suicide, sex, substance abuse, and other potential pitfalls. Few of them work.

Yet we can and do help. We can ward off some of the world's worst hazards and nudge adolescents toward appropriate responses to the rest. Studies show that when parents engage and guide their teens with a light but steady hand, staying connected but allowing independence, their kids generally do much better in life. Adolescents want to learn primarily, but not entirely, from their friends. At some level and at some times (and it's the parent's job to spot when), the teen recognizes that the parent can offer certain kernels of wisdom—knowledge valued not because it comes from parental authority but because it comes from the parent's own struggles to learn how the world turns. The teen rightly perceives that she must understand not just her parents' world but also the one she is entering. Yet if allowed to, she can appreciate that her parents once faced the same problems and may remember a few things worth knowing.

MEANWHILE, IN TIMES OF DOUBT, take inspiration in one last distinction of the teen brain—a final key to both its clumsiness and its remarkable adaptability. This is the prolonged plasticity of those late-developing frontal areas as they slowly mature. As noted earlier, these areas are the last to lay down the fatty myelin insulation—the brain's white matter—that speeds transmission. And at first glance this seems like bad news: If we need these areas for the complex task of entering the world, why aren't they running at full speed when the challenges are most daunting?

The answer is that speed comes at the price of flexibility. While a myelin coating greatly accelerates an axon's bandwidth, it also in-hibits the growth of new branches from the axon. According to Douglas Fields, an NIH neuroscientist who has spent years studying myelin, "This makes the period when a brain area lays down myelin a sort of crucial period of learning—the wiring is getting upgraded, but once that's done, it's harder to change."

The window in which experience can best rewire those connec-tions is highly specific to each brain area. Thus the brain's language centers acquire their insulation most heavily in the first 13 years, when a child is learning language. The completed insulation con-solidates those gains—but makes further gains, such as second lan-guages, far harder to come by.

So it is with the forebrain's myelination during the late teens and early 20s. This delayed completion—a withholding of readiness—heightens flexibility just as we confront and enter the world that we will face as adults.

This long, slow, back-to-front developmental wave, completed only in the mid-20s, appears to be a uniquely human adaptation. It may be one of our most consequential. It can seem a bit crazy that we humans don't wise up a bit earlier in life. But if we smartened up sooner, we'd end up dumber.

JOSH FISCHMAN

Criminal Minds

FROM THE *CHRONICLE OF HIGHER EDUCATION*

> There is now evidence that connects the size and nature of certain
> features of the brain to criminal behavior. Josh Fischman looks into
> whether this means we will be able to predict which children will
> grow up to be killers.

HE WAS LOCKED IN A VAN IN ENGLAND WITH VIO-
lent criminals, repeatedly, during his late 20s, says Adrian
Raine, lifting a fork of salmon ravioli from his plate at a tony
restaurant on Walnut Street. "I was at the maximum-security prison
in Hull," says the psychologist, now in his 50s, and his job involved
attaching polygraph-type sensors to the prisoners' skin to measure
their agitation as he bothered them with loud sounds and flashes of
light. His lab was in the back of the van, he says, "and the guards were
very concerned these men would commandeer the vehicle and escape."

Their solution? "Take my keys away and lock the doors from the outside."

Raine, now chair of the criminology department at the University of Pennsylvania, a few blocks from the restaurant, stops eating for a moment to remember. "So there I was, in this very tiny space. And I kept watching the needles these sensors were connected to, for I imagined that the first sign these men were about to rush me would be the needles starting to swing wildly as the men got excited and prepared to attack."

They never did. Raine always got out of the van unscathed, but the slightly built, graying Englishman has never strayed far from the company of killers, wife batterers, and psychopaths. He has spent a career trying to spot ever-earlier signs of dangerous minds—clues to bad behavior even before a criminal commits a crime.

Along with several other researchers, he has pioneered the science of neurodevelopmental criminology. In adult offenders, juvenile delinquents, and even younger children, dozens of studies have pointed to brain features that seem to reduce fear, impair decision making, and blunt emotional reactions to others' distress. The studies have also highlighted body reactions that are signs of this pattern and are tied to criminality.

Society has always wondered about "bad seeds," people who seem to be possessed by devils. But what is emerging from this research is a cluster of biological markers that plant the bad seed in the brain. More striking, they appear to predict antisocial behavior even before it happens. Early warnings could avoid a world of hurt, because some of these people are terribly dangerous. Kent A. Kiehl, an associate professor of psychology at the University of New Mexico, has described in The New Yorker a psychopath whom he encountered during his research: The man started with petty crimes as a child and was convicted of arson by age 17. After a prison stretch, he moved back in with his mother, got in an argument with her, and wrapped the phone cord around her neck when she tried to call for help. He then threw her down some stairs, stabbed her several times, crushed her head beneath a propane canister, and went out to celebrate with a three-day bender.

"So if I could tell you, as a parent, that your child has a 75-percent chance of becoming a criminal, wouldn't you want to know and maybe have the chance to do something about it?" asks Raine.

But predicting criminality makes a lot of people nervous. It raises the specter of earlier scientific episodes in which researchers claimed that aspects of biology—like ethnic background—determined behavior or intelligence. With what everyone admits is still a young and imperfect science, there is also the possibility of getting it wrong, and ruining someone's future with an undeserved "bad seed" label.

Still, Raine has many supporters. "He really has a broad range of work," says Abigail A. Marsh, an assistant professor of psychology at Georgetown University who has studied brain areas related to psychopathy. "I'm particularly impressed with his long-term studies, where he's traced physical patterns in children to conduct problems in adulthood, because doing that kind of research is difficult."

Raine himself acknowledges that he's going down a dangerous path. "The brain changes. Family and the environment may protect it and may even alter some of the physiology that we see," he says. "So we don't want to say that biology is destiny." Indeed, he is in the midst of an experiment with young children who show some markers of aggressive behavior, to see if psychotherapy or a diet rich in brain-protective supplements can reduce chances of antisocial acts.

"Of course, all of this brings up tremendously difficult ethical questions," he says. "But I don't think I'd be doing my job unless I said that we need to start talking about them. It's time we start this discussion, before we start labeling people."

THE NOTION OF A CHANGEABLE brain was not at all part of the last serious scientific attempt to link physical traits to behavior or thinking. In the 19th century, physical anthropologists like Paul Broca and Paul Topinard—founding figures in the field—took calipers to skulls and scales to brains to show that specific brain shapes were features of particular races or groups, and that the mental abilities of those races followed those shapes almost without fail. There was the "facial angle," a degree of forward thrusting that was supposed to mark intelligence. Europeans purportedly had an angle of 80 degrees, Africans of 70 degrees, and apes rather less. The method was called craniometry, and it served an assumption that racial differences were primordial and showed up as immutable mental characteristics. (Phrenology, a pseudoscience that tied various bumps on the head to fine-grained personality traits, was never taken seriously by researchers.) The notion began to unravel when other scientists, like Franz Boas in the early 20th century, took up their own calipers to show that the anatomy of a race in fact shifted across generations.

And as Raine points out, the individual brain does change. Beginning in the 1960s, scientists began amassing evidence of "plasticity" when they found brain regions taking on new functions to make up for injury. In the 1980s and 1990s, they established that animals generate new neurons when they learn new skills. They also found that human babies reshape their brain pathways as they grow. Anatomy at birth may influence many things, but it is not fate.

It was in that environment that Raine began exploring psychology. He was born the son of a bricklayer and an ice-cream vendor in Darlington, in the north of England. During a brief flirtation with an accountant's career ("I was a boring kid and good with numbers," he says), he was wandering in a bookstore one weekend afternoon. "I came across a book by psychiatrist R.D. Laing on how faulty family communication can make you a schizophrenic. It was like an epiphany for me, reading about this effect on brain and behavior. So then I wanted to go to university and be a psychiatrist."

The university turned out to be Oxford—Jesus College—where

Raine vacillated between wanting to be an experimental psychologist and wanting to teach primary school. That second impulse, though not followed through, was fateful. Student teaching "showed me there were some kids who were just bullies, very extreme, and I wondered, Why? Where did the behavior come from? Why were some kids angels and some devils?"

He made the topic into his doctoral research at the University of York, where he began using the lie-detector-like sensors, later featured in his prison van, to measure heart rates and changes in the way the skin conducts electricity in teenagers with varying degrees of aggression.

But after he got his degree, he couldn't get a job. "Biological explanations were not popular at that time, the late 1970s. It was not on anyone's radar screen at all. I sent out 67 job applications, to universities as far away as Papua New Guinea, and I got 67 rejections. The only place that would take me was prison."

As a prison psychologist at Hull, starting in 1980, he worked with rapists, murderers, and pedophiles. That was also the year that Robert D. Hare, now an emeritus professor at the University of British Columbia, published the first checklist that could single out psychopaths from other personality types, and Raine began trying various interventions with that group. "The main thing that I learned was that I really can't change these people," he says. "It brought me back to kids, to earlier stages, thinking we've got to look at the earlier, predisposing factors, when maybe we can do something."

After two years at Hull and two at another prison, Raine broke out of jail and into academe, first at the University of Nottingham and then at the University of Southern California, always pursuing those predispositions. One of his major lines of research was long-term studies of children, measuring physiological reactivity at young ages to see if any pattern related to bad behavior decades later.

Testing and following children for years is not, as Georgetown's Abigail Marsh says, a simple thing to do. They tend to drop out, move away, disappear. But Raine lucked into a major source of data. In the late 1960s, the World Health Organization had started a project to closely follow about 1,800 children on Mauritius, a small island nation in the Indian Ocean. The nice thing about remote islands: People don't tend to move away and disappear.

Over the years, Raine and several colleagues have shown that children from Mauritius who show slower heart rates and reduced skin responses when annoyed by loud tones or challenging questions tend to have criminal records when they get older. In 1996 the researchers showed that 15-year-olds with this pattern tended to have criminal records by age 29. In 2010, the age was pushed back further: 3-year-olds who had those physical responses were rated by teachers as more aggressive than other children five years later.

But when such kids got better schooling and nutrition starting at age 3, they had more typical physical reactions. And by age 23, their incidence of criminal behavior dropped by 35 percent.

There is a theory behind this, and it's about being insensitive to fear. Normally, a startling noise races the heart and sends the body into a high state of alert, which is what the skin electrodes pick up. But research indicates that children who are not alarmed don't react to the threat of punishment when they misbehave. Nor do they react to the distress shown by other people. They don't learn that their bad actions, like causing others pain, have bad consequences for those people. The pattern builds on itself until—maybe—it creates a person who wraps a telephone cord around his mother's throat.

Another line of evidence supporting this idea—the cold-blooded criminal—comes from studies of young twins who have what are called callous, unemotional traits. At Southern California, Raine was part of a team studying 605 families of twins. Some of the twins were identical, sharing 100 percent of their genes, and some were nonidentical, sharing 50 percent. When identical twins show more similar traits than nonidentical pairs, it most likely means that genes have a stronger influence than environment, because all of the twins were raised in the same surroundings.

In Raine's twins studies, when the children were ages 9 and 10, researchers gave them a battery of psychological tests to assess aggression and antisocial behavior. They also asked three different kinds of observers—parents, teachers, and the children themselves—to rate the children's antisocial tendencies: The results showed that these traits were more consistent in the identical pairs. "I think at least 50 percent of this can be attributed to genetics," Raine says.

More recently, Nathalie M.G. Fontaine, an assistant professor of criminology at Indiana University at Bloomington, and a team of researchers have followed more than 9,000 sets of twins in England and Wales from the time they were 4 through age 12. Those young children in whom parents and teachers saw repeated callous behavior and lack of emotion were the ones most likely to have patterns of lying, cheating, and stealing by age 12. Moreover, "biology appears to play a role in this," Fontaine says, because traits appear to be carried across twins. "In boys with high CU [callous, unemotional] traits, for example, we think 78 percent of their CU traits are inherited," she said when she presented the findings at this year's meeting of the American Association for the Advancement of Science.

After years of observing behavior like this, and noting its repeated connection to dulled nerves, Raine began wondering what blunted the senses. Ultimately, of course, all nerves connect to the brain. So was there something actually in the brains of these people that tied into this lack of fear, lack of sensitivity, and abundance of mayhem?

Others had been considering the same thing. Antonio R. Damasio, a pioneering professor of neuroscience, was one of the first to demonstrate that damage to parts of the brain is tied to antisocial actions, in work done during the 1990s at the University of Iowa. (He is now at Southern California.) He focused on the amygdala, a small, almond-shaped area in the middle of the brain known to help process emotions. In patients with lesions in the region, he observed a repeated pattern of bad decisions, like making risky bets while gambling. Cut off from emotional reactions, these people lacked an alarm bell that signaled a poor choice. They could see many courses of action but, shorn of feedback from the amygdala, couldn't tell good from bad. That extended to reading other people's emotions, too: The patients had trouble distinguishing frightened expressions from happy ones. James R. Blair, a neuroscientist at the National Institute of Mental Health, has found these abnormalities to be especially common among psychopaths, which could explain their unsettling ability to hurt other people horribly yet feel no remorse at all.

Damasio also found that damage to the prefrontal cortex, a brain area involved in decision making, could turn mild-mannered patients into rash, destructive individuals, seeming to rob them of a "brake" on their impulses. This led him to propose that the two regions normally link up to prevent people from harming others by generating emotional alarm (the amygdala) and acting on it (the prefrontal cortex). When either part of the chain is damaged, antisocial actions result. Damasio called this "acquired sociopathy."

In 1997, Raine and several colleagues put the theory to the test on real killers. They compared the functioning of the brains of 41 convicted murderers with those of 41 normal people. Using positron emission tomography, or PET, a type of scan that measures the activity in areas of the brain, they saw lower activity in both the prefrontal cortex and the amygdala of the murderers' brains.

When they further divided murderers into those who came from "good" homes and those who came from "bad" homes—those filled with neglect, abuse, and poverty—the first group again showed lower activity in the prefrontal cortex, in particular an area called the orbitofrontal cortex. Raine's interpretation: Genetics and anatomy were more influential on their development than was the way they grew up; the murderers from good homes seemed to be terribly affected by this low-functioning brain region.

And it wasn't just function. Brain form was also impaired, Raine and his coworkers found. A series of studies using magnetic resonance imaging, which reveals structures and shapes, showed that criminals and people who scored high on tests of antisocial disorders had different-looking brains. Both the orbitofrontal region and the amygdala were smaller than normal. And the corpus callosum, the bridge between the brain's two hemispheres that helps them communicate, was abnormally large.

But those findings raised a chicken-and-egg type question: Did the brain features produce the behavior, or did the behavior change the brain? Violent criminals are known to bang their heads into walls and abuse drugs, and both of those things damage the brain, possibly producing the shrinkage Raine was seeing on the brain scans. He needed to go back even further, to birth and beyond.

You can't put a fetus in a brain scanner. What you can do, however, is look for a defect that begins before birth and can still be detected in adults. Raine found it in a hole in the head. More precisely, a thin wall of brain tissue that separates a hole—all brains have these spaces—into two. The hole appears during the 12th week of a fetus's development, and the wall—pushed forward by a normally developing amygdala and other brain areas—divides it by the 20th week. When the wall doesn't form completely, a condition known by the jawbreaking name of cavum septum pellucidum, it's usually a sign of abnormal development in the amygdala and other structures. Years later, in adults, the failed wall can be spotted in a brain scan.

Raine, who moved in 2007 to the University of Pennsylvania, found that the condition is also associated with dangerous minds. In a 2010 paper, he and his colleagues compared people with and without the feature on several fronts. The groups were tested for antisocial personality disorder, psychopathy, and aggression. Their records were searched for criminal arrests and convictions. In every single one of those areas, there were a lot more men and women with the wall defect. Here, finally, was evidence tracing criminality back to the womb, before any head-banging could occur. It was evidence of changes in the brain that nicely tied in with the abnormalities seen in adult criminals.

"I think there's no longer any question, scientifically, that there's an association between the brain and criminal behavior. We're beyond the point of debating that," says Raine. "Every study can be criticized on methodology. But when you look at the whole, at all the different designs, it's just hard to deny there is something going on with biology."

Indeed, the evidence keeps coming: This month a team of researchers from the Universities of Cambridge and of Southampton reported that 15-year-old kids with conduct disorders had smaller amygdalae than those with no behavior problems; the more severe the disorder, the greater the shrinkage. Some genes have even been linked, tentatively, to these conduct disorders.

So what do we do now?

One thing we don't do, says Fontaine, is mark children as future criminals. "No way! I would never put a risk number on a specific child," she says. "We are talking about groups, not individuals. We don't know what will happen with any one child, because there are also protective factors." The plastic, changeable brain means that a strongly supportive family, or an influential schoolteacher, or religion could blunt the effects of a hole in the head. Or as Damasio, who thinks Raine has produced a strong body of evidence, puts it: "The changes are themselves changeable."

Raine himself is a big believer in protective factors. "You can't make a lesion to the prefrontal cortex and, hey presto, you get a criminal. It's not like that," he says. "Of course social factors are critically important." In his current study of Philadelphia children with the slow physical reactivity that has been linked to trouble, some are getting a diet rich in omega-3 fatty acids and calcium to see if those protect brain cells, some are getting cognitive-behavioral therapy, and some are getting both to see if trouble can be staved off.

Still, the time is coming, Raine believes, when putting numbers on children will be tempting. If a 75-percent chance of a bad seed isn't high enough, he wonders, what about 80 percent? Or 95? "Look, I have two children, 9-year-old, nonidentical twin boys," he says. "And I'd definitely want to know, especially if there was a treatment that has a chance of success. But I realize not every parent will. We have to start having this conversation now, though, so we understand the risks and the benefits. It's easy to get on your moral high horse about stigma and civil liberties, but are you going to have blood on your hands in the future because you've blocked an approach that could lead to lives being saved?

"One swallow does not a summer make. But together, this is a message in the sky."

DOUGLAS FOX

The Limits of Intelligence

FROM *SCIENTIFIC AMERICAN*

> When it comes to brains, bigger may not be better: the nature of a
> brain's wiring, and not its size, may be a better predictor of intel-
> ligence. This may imply, as Douglas Fox learns, a natural limit to
> how smart we can get.

SANTIAGO RAMÓ Y CAJAL, THE SPANISH NOBEL-
winning biologist who mapped the neural anatomy of insects
in the decades before World War I, likened the minute circuitry
of their vision-processing neurons to an exquisite pocket watch. He
likened that of mammals, by comparison, to a hollow-chested
grandfather clock. Indeed, it is humbling to think that a honeybee,
with its milligram-size brain, can perform tasks such as navigating
mazes and landscapes on a par with mammals. A honeybee may be

limited by having comparatively few neurons, but it surely seems to squeeze everything it can out of them.

At the other extreme, an elephant, with its five-million-fold larger brain, suffers the inefficiencies of a sprawling Mesopotamian empire. Signals take more than 100 times longer to travel between opposite sides of its brain—and also from its brain to its foot, forcing the beast to rely less on reflexes, to move more slowly, and to squander precious brain resources on planning each step.

We humans may not occupy the dimensional extremes of elephants or honeybees, but what few people realize is that the laws of physics place tough constraints on our mental faculties as well. Anthropologists have speculated about anatomic roadblocks to brain expansion—for instance, whether a larger brain could fit through the birth canal of a bipedal human. If we assume, though, that evolution can solve the birth canal problem, then we are led to the cusp of some even more profound questions.

One might think, for example, that evolutionary processes could increase the number of neurons in our brain or boost the rate at which those neurons exchange information and that such changes would make us smarter. But several recent trends of investigation, if taken together and followed to their logical conclusion, seem to suggest that such tweaks would soon run into physical limits. Ultimately those limits trace back to the very nature of neurons and the statistically noisy chemical exchanges by which they communicate. "Information, noise and energy are inextricably linked," says Simon Laughlin, a theoretical neuroscientist at the University of Cambridge. "That connection exists at the thermodynamic level."

Do the laws of thermodynamics, then, impose a limit on neuron-based intelligence, one that applies universally, whether in birds, primates, porpoises or praying mantises? This question apparently has never been asked in such broad terms, but the scientists interviewed for this article generally agree that it is a question worth contemplating. "It's a very interesting point," says Vijay Balasubramanian, a physicist who studies neural coding of information at the University of Pennsylvania. "I've never even seen this point discussed in science fiction."

Intelligence is of course a loaded word: it is hard to measure and even to define. Still, it seems fair to say that by most metrics, humans are the most intelligent animals on earth. But as our brain has evolved, has it approached a hard limit to its ability to process information? Could there be some physical limit to the evolution of neuron-based intelligence—and not just for humans but for all of life as we know it?

THAT HUNGRY TAPEWORM IN YOUR HEAD

The most intuitively obvious way in which brains could get more powerful is by growing larger. And indeed, the possible connection between brain size and intelligence has fascinated scientists for more than 100 years. Biologists spent much of the late 19th century and the early 20th century exploring universal themes of life— mathematical laws related to body mass, and to brain mass in particular, that run across the animal kingdom. One advantage of size is that a larger brain can contain more neurons, which should enable it to grow in complexity as well. But it was clear even then that brain size alone did not determine intelligence: a cow carries a brain well over 100 times larger than a mouse's, but the cow isn't any smarter. Instead brains seem to expand with body size to carry out more trivial functions: bigger bodies might, for example, impose a larger workload of neural housekeeping chores unrelated to intelligence, such as monitoring more tactile nerves, processing signals from larger retinas and controlling more muscle fibers.

Eugene Dubois, the Dutch anatomist who discovered the skull of Homo erectus in Java in 1892, wanted a way to estimate the intelligence of animals based on the size of their fossil skulls, so he worked to define a precise mathematical relation between the brain size and body size of animals—under the assumption that animals with disproportionately large brains would also be smarter. Dubois and others amassed an ever growing database of brain and body weights; one classic treatise reported the body, organ and gland weights of 3,690 animals, from wood roaches to yellow-billed egrets to two-toed and three-toed sloths.

Dubois's successors found that mammals' brains expand more slowly than their bodies—to about the ¾ power of body mass. So a muskrat, with a body 16 times larger than a mouse's, has a brain about eight times as big. From that insight came the tool that Dubois had sought: the encephalization quotient, which compares a species' brain mass with what is predicted based on body mass. In other words, it indicates by what factor a species deviates from the ¾ power law. Humans have a quotient of 7.5 (our brain is 7.5 times larger than the law predicts); bottlenose dolphins sit at 5.3; monkeys hover as high as 4.8; and oxen—no surprise there—slink around at 0.5. In short, intelligence may depend on the amount of neural reserve that is left over after the brain's menial chores, such as minding skin sensations, are accounted for. Or to boil it down even more: intelligence may depend on brain size in at least a superficial way.

As brains expanded in mammals and birds, they almost certainly benefited from economies of scale. For example, the greater number of neural pathways that any one signal between neurons can travel means that each signal implicitly carries more information, implying that the neurons in larger brains can get away with firing fewer times per second. Meanwhile, however, another, competing trend may have kicked in. "I think it is very likely that there is a law of diminishing returns" to increasing intelligence indefinitely by adding new brain cells, Balasubramanian says. Size carries burdens with it, the most obvious one being added energy consumption. In humans, the brain is already the hungriest part of our body: at 2 percent of our body weight, this greedy little tapeworm of an organ wolfs down 20 percent of the calories that we expend at rest. In newborns, it's an astounding 65 percent.

Staying in Touch

Much of the energetic burden of brain size comes from the organ's communication networks: in the human cortex, communications account for 80 percent of energy consumption. But it appears that as size increases, neuronal connectivity also becomes more challenging for subtler, structural reasons. In fact, even as biologists kept collecting data on brain mass in the early to mid-20th century, they delved into a more daunting enterprise: to define the "design principles" of brains and how these principles are maintained across brains of different sizes.

A typical neuron has an elongated tail called the axon. At its end, the axon branches out, with the tips of the branches forming synapses, or contact points, with other cells. Axons, like telegraph wires, may connect different parts of the brain or may bundle up into nerves that extend from the central nervous system to the various parts of the body.

In their pioneering efforts, biologists measured the diameter of axons under microscopes and counted the size and density of nerve cells and the number of synapses per cell. They surveyed hundreds, sometimes thousands, of cells per brain in dozens of species. Eager to refine their mathematical curves by extending them to ever larger beasts, they even found ways to extract intact brains from whale carcasses. The five-hour process, meticulously described in the 1880s by biologist Gustav Adolf Guldberg, involved the use of a two-man lumberjack saw, an ax, a chisel and plenty of strength to open the top of the skull like a can of beans.

These studies revealed that as brains expand in size from species to species, several subtle but probably unsustainable changes happen. First, the average size of nerve cells increases. This phenomenon allows the neurons to connect to more and more of their compatriots as the overall number of neurons in the brain increases. But larger cells pack into the cerebral cortex less densely, so the distance between cells increases, as does the length of axons required to connect them. And because longer axons mean longer times for signals to travel between cells, these projections need to become thicker to maintain speed (thicker axons carry signals faster).

Researchers have also found that as brains get bigger from species to species, they are divided into a larger and larger number of distinct areas. You can see those areas if you stain brain tissue and view it under a microscope: patches of the cortex turn different colors. These areas often correspond with specialized functions, say, speech comprehension or face recognition. And as brains get larger, the specialization unfolds in another dimension: equivalent areas in the left and right hemispheres take on separate functions—for example, spatial versus verbal reasoning.

For decades this dividing of the brain into more work cubicles was viewed as a hallmark of intelligence. But it may also reflect a more mundane truth, says Mark Changizi, a theoretical neurobiologist at 2AI Labs in Boise, Idaho: specialization compensates for the connectivity problem that arises as brains get bigger. As you go from a mouse brain to a cow brain with 100 times as many neurons, it is impossible for neurons to expand quickly enough to stay just as well connected. Brains solve this problem by segregating like-functioned neurons into highly interconnected modules, with far fewer long-distance connections between modules. The specialization between right and left hemispheres solves a similar problem: it reduces the amount of information that must flow between the hemispheres, which minimizes the number of long, interhemispheric axons that the brain needs to maintain. "All of these seemingly complex things about bigger brains are just the backbends that the brain has to do to satisfy the connectivity problem" as it gets larger, Changizi argues. "It doesn't tell us that the brain is smarter."

Jan Karbowski, a computational neuroscientist at the Polish Academy of Sciences in Warsaw, agrees. "Somehow brains have to optimize several parameters simultaneously, and there must be trade-offs," he says. "If you want to improve one thing, you screw up something else." What happens, for example, if you expand the corpus callosum (the bundle of axons connecting right and left hemispheres) quickly enough to maintain constant connectivity as brains expand? And what if you thicken those axons, so the transit delay for signals traveling between hemispheres does not increase as brains expand? The results would not be pretty. The corpus callosum would expand—and push the hemispheres apart—so quickly that any performance improvements would be neutralized.

These trade-offs have been laid into stark relief by experiments showing the relation between axon width and conduction speed. At the end of the day, Karbowski says, neurons do get larger as brain size increases, but not quite quickly enough to stay equally well connected. And axons do get thicker as brains expand, but not quickly enough to make up for the longer conduction delays.

Keeping axons from thickening too quickly saves not only space but energy as well, Balasubramanian says. Doubling the width of an axon doubles energy expenditure, while increasing the velocity of pulses by just 40 percent or so. Even with all of this corner cutting, the volume of white matter (the axons) still grows more quickly than the volume of gray matter (the main body of neurons containing the cell nucleus) as brains increase in size. To put it another way, as brains get bigger, more of their volume is devoted to wiring rather than to the parts of individual cells that do the actual computing, which again suggests that scaling size up is ultimately unsustainable.

THE PRIMACY OF PRIMATES

It is easy, with this dire state of affairs, to see why a cow fails to squeeze any more smarts out of its grapefruit-size brain than a mouse does from its blueberry-size brain. But evolution has also achieved impressive workarounds at the level of the brain's building blocks. When Jon H. Kaas, a neuroscientist at Vanderbilt University, and his colleagues compared the morphology of brain cells across a spectrum of primates in 2007, they stumbled onto a game changer—one that has probably given humans an edge.

Kaas found that unlike in most other mammals, cortical neurons in primates enlarge very little as the brain increases in size. A few neurons do increase in size, and these rare ones may shoulder the burden of keeping things well connected. But the majority do not get larger. Thus, as primate brains expand from species to species, their neurons still pack together almost as densely. So from the marmoset to the owl monkey—a doubling in brain mass—the number of neurons roughly doubles, whereas in rodents with a similar doubling of mass the number of neurons increases by just 60 percent. That difference has huge consequences. Humans pack 100 billion neurons into 1.4 kilograms of brain, but a rodent that had followed its usual neuron-size scaling law to reach that number of neurons would now have to drag around a brain weighing 45 kilograms. And metabolically speaking, all that brain matter would eat the varmint out of house and home. "That may be one of the factors in why the large rodents don't seem to be [smarter] at all than the small rodents," Kaas says.

Having smaller, more densely packed neurons does seem to have a real impact on intelligence. In 2005 neurobiologists Gerhard Roth and Urusula Dicke, both at the University of Bremen in Germany, reviewed several traits that predict intelligence across species (as measured, roughly, by behavioral complexity) even more effectively than the encephalization quotient does. "The only tight correlation with intelligence," Roth says, "is in the number of neurons in the cortex, plus the speed of neuronal activity," which decreases with the distance between neurons and increases with the degree of myelination of axons. Myelin is fatty insulation that lets axons transmit signals more quickly.

If Roth is right, then primates' small neurons have a double effect: first, they allow a greater increase in cortical cell number as brains enlarge; and second, they allow faster communication, because the cells pack more closely. Elephants and whales are reasonably smart, but their larger neurons and bigger brains lead to inefficiencies. "The packing density of neurons is much lower," Roth says, "which means that the distance between neurons is larger and the velocity of nerve impulses is much lower."

In fact, neuroscientists have recently seen a similar pattern in variations within humans: people with the quickest lines of communication between their brain areas also seem to be the brightest. One study, led in 2009 by Martijn P. van den Heuvel of the University Medical Center Utrecht in the Netherlands, used functional magnetic resonance imaging to measure how directly different brain areas talk to one another—that is, whether they talk via a large or a small number of intermediary areas. Van den Heuvel found that shorter paths between brain areas correlated with higher IQ. Edward Bullmore, an imaging neuroscientist at the University of Cambridge, and his collaborators obtained similar results the same year using a different approach. They compared working memory (the ability to hold several numbers in one's memory at once) among 29 healthy people. They then used magnetoencephalographic recordings from their subjects' scalp to estimate how quickly communication flowed between brain areas. People with the most direct communication and the fastest neural chatter had the best working memory.

It is a momentous insight. We know that as brains get larger, they save space and energy by limiting the number of direct connections between regions. The large human brain has relatively few of these long-distance connections. But Bullmore and van den Heuvel showed that these rare, nonstop connections have a disproportionate influence on smarts: brains that scrimp on resources by cutting just a few of them do noticeably worse. "You pay a price for intelligence," Bullmore concludes, "and the price is that you can't simply minimize wiring."

INTELLIGENCE DESIGN

If communication between neurons, and between brain areas, is really a major bottleneck that limits intelligence, then evolving neurons that are even smaller (and closer together, with faster communication) should yield smarter brains. Similarly, brains might become more efficient by evolving axons that can carry signals faster over longer distances without getting thicker. But something prevents animals from shrinking neurons and axons beyond a certain point. You might call it the mother of all limitations: the proteins that neurons use to generate electrical pulses, called ion channels, are inherently unreliable.

Ion channels are tiny valves that open and close through changes in their molecular folding. When they open, they allow ions of sodium, potassium or calcium to flow across cell membranes, producing the electrical signals by which neurons communicate. But being so minuscule, ion channels can get flipped open or closed by mere thermal vibrations. A simple biology experiment lays the defect bare. Isolate a single ion channel on the surface of a nerve cell using a microscopic glass tube, sort of like slipping a glass cup over a single ant on a sidewalk. When you adjust the voltage on the ion channel— a maneuver that causes it to open or close—the ion channel does not flip on and off reliably like your kitchen light does. Instead it flutters on and off randomly. Sometimes it does not open at all; other times it opens when it should not. By changing the voltage, all you do is change the likelihood that it opens.

It sounds like a horrible evolutionary design flaw—but in fact, it is a compromise. "If you make the spring on the channel too loose, then the noise keeps on switching it," Laughlin says—as happens in the biology experiment described earlier. "If you make the spring on the channel stronger, then you get less noise," he says, "but now it's more work to switch it," which forces neurons to spend more energy to control the ion channel. In other words, neurons save energy by using hair-trigger ion channels, but as a side effect the channels can flip open or close accidentally. The trade-off means that ion channels are reliable only if you use large numbers of them to "vote" on whether or not a neuron will generate an impulse. But voting becomes problematic as neurons get smaller. "When you reduce the size of neurons, you reduce the number of channels that are available to carry the signal," Laughlin says. "And that increases the noise."

In a pair of papers published in 2005 and 2007, Laughlin and his collaborators calculated whether the need to include enough ion channels limits how small axons can be made. The results were startling. "When axons got to be about 150 to 200 nanometers in diameter, they became impossibly noisy," Laughlin says. At that point, an axon contains so few ion channels that the accidental opening of a single channel can spur the axon to deliver a signal even though the neuron did not intend to fire. The brain's smallest axons probably already hiccup out about six of these accidental spikes per second. Shrink them just a little bit more, and they would blather out more than 100 per second. "Cortical gray matter neurons are working with axons that are pretty close to the physical limit," Laughlin concludes.

This fundamental compromise between information, energy and noise is not unique to biology. It applies to everything from optical-fiber communications to ham radios and computer chips. Transistors act as gatekeepers of electrical signals, just like ion channels do. For five decades engineers have shrunk transistors steadily, cramming more and more onto chips to produce ever faster computers. Transistors in the latest chips are 22 nanometers. At those sizes, it becomes very challenging to "dope" silicon uniformly (doping is the addition of small quantities of other elements to adjust a semiconductor's properties). By the time they reach about 10 nanometers, transistors will be so small that the random presence or absence of a single atom of boron will cause them to behave unpredictably.

Engineers might circumvent the limitations of current transistors by going back to the drawing board and redesigning chips to use entirely new technologies. But evolution cannot start from scratch: it has to work within the scheme and with the parts that have existed for half a billion years, explains Heinrich Reichert, a developmental neurobiologist at the University of Basel in Switzerland—like building a battleship with modified airplane parts.

Moreover, there is another reason to doubt that a major evolutionary leap could lead to smarter brains. Biology may have had a wide range of options when neurons first evolved, but 600 million years later a peculiar thing has happened. The brains of the honeybee, the octopus, the crow and intelligent mammals, Roth points out, look nothing alike at first glance. But if you look at the circuits that underlie tasks such as vision, smell, navigation and episodic memory of event sequences, "very astonishingly they all have absolutely the same basic arrangement." Such evolutionary convergence usually suggests that a certain anatomical or physiological solution has reached maturity so that there may be little room left for improvement.

Perhaps, then, life has arrived at an optimal neural blueprint. That blueprint is wired up through a step-by-step choreography in which cells in the growing embryo interact through signaling molecules and physical nudging, and it is evolutionarily entrenched.

BEES DO IT

So have humans reached the physical limits of how complex our brain can be, given the building blocks that are available to us? Laughlin doubts that there is any hard limit on brain function the way there is one on the speed of light. "It's more likely you just have a law of diminishing returns," he says. "It becomes less and less worthwhile the more you invest in it." Our brain can pack in only so many neurons; our neurons can establish only so many connections among themselves; and those connections can carry only so many electrical impulses per second. Moreover, if our body and brain got much bigger, there would be costs in terms of energy consumption, dissipation of heat and the sheer time it takes for neural impulses to travel from one part of the brain to another.

The human mind, however, may have better ways of expanding without the need for further biological evolution. After all, honeybees and other social insects do it: acting in concert with their hive sisters, they form a collective entity that is smarter than the sum of its parts. Through social interaction we, too, have learned to pool our intelligence with others.

And then there is technology. For millennia written language has enabled us to store information outside our body, beyond the capacity of our brain to memorize. One could argue that the Internet is the ultimate consequence of this trend toward outward expansion of intelligence beyond our body. In a sense, it could be true, as some say, that the Internet makes you stupid: collective human intelligence—culture and computers—may have reduced the impetus for evolving greater individual smarts.

JARON LANIER

It's Not a Game

FROM *TECHNOLOGY REVIEW*

Many celebrated the triumph of IBM's Watson computer over its human competitors on Jeopardy! *as heralding a new era in artificial intelligence. Jaron Lanier has his doubts.*

WATCHING THE COMPUTER SYSTEM KNOWN AS Watson defeat the top two human *Jeopardy!* players of all time was fun in the short term. This demonstration of IBM's software, however, was a bad idea in the longer term. It presented a misleading picture to the public of what is known about machine and human intelligence, and more seriously, it advanced a flawed approach to science that stands to benefit the enemies of science.

There's a crucial distinction to make right away. My purpose is not to criticize the work done by the team that created Watson. Nor do I want to critique their professional publications or their interactions with colleagues in the field of computer science. Instead, I am concerned with the nature of the pop spectacle hatched by IBM.

Why was there a public spectacle at all? Certainly it's worthwhile to share the joy and excitement of science with the public, as NASA often does. But there were no other Mars rovers to compare with the NASA rovers when they landed, and there is a whole world of research related to artificial intelligence. By putting its system on TV and personifying that system with a name and a computer-generated voice, IBM separated it from its context, suggesting—falsely—the existence of a sui generis entity.

Contrast IBM's theatrics with the introduction of Wolfram Alpha, a "knowledge engine" for the Web that physicist Stephen Wolfram released in 2009. Although the early rhetoric around Alpha was a touch extreme, sometimes exaggerating its natural-language competence, the method of introduction was vastly more honest. Wolfram Research didn't resort to stage magic: Alpha was made available online for people to try. Stephen Wolfram encouraged people to use his technology and compare the results with those generated by search engines like Google. Alpha proved honestly that it was something fresh, different, and useful. Comparison with what came before is crucial to progress in science and technology.

But Watson was presented on TV as an entity instead of a technology, and people are inclined to treat entities charitably. You are more likely to give a "he" the benefit of a doubt, while you judge an "it" for what it can do as a tool. Watson avoided any such comparative judgment, and the public wasn't given a window into what would happen in that kind of empirical process. Stephen Wolfram himself, however, went to the trouble of writing a blog post comparing Watson with everyday search engines. He entered the text of *Jeopardy!* clues into those search engines and found that in many cases, the first document they returned contained the answer. Identifying a page that contains the answer is not the same thing as being able to give the answer on *Jeopardy!*, but this little experiment does indicate that Watson's abilities were less extraordinary than one might have gathered from watching the broadcast.

Wouldn't it have been better to open the legitimate process of science to the public instead of staging a fake version? An example of how to do this was the DARPA-sponsored "Grand Challenge" to create self-driving cars. By pitting technologies against each other, DARPA informed the public well and offered a glimpse into the state of the art. The contest also made for great TV. Competitors were motivated. The process worked.

The *Jeopardy!* show in itself, by contrast, was not informative. There are a multitude of open questions about how human language works and how brains think. But when machines are pitted against people, an unstated assertion is inevitably propagated: that human thinking and machine "intelligence" are already known to be at least comparable. Of course, this is not true.

In the case of *Jeopardy!*, the game's design isolates a specific skill: guessing words on the basis of hints. We know that being able to guess an unstated word from its context is part of language competency, but we don't know how important that skill is in relation to the whole phenomenon of human language. We don't fully know what would be required to re-create that phenomenon. Even if it had been stated (in fine print, as it were) that the task of competing at *Jeopardy!* shouldn't be confused with complete mastery of human language, the extravaganza would have left the impression that scientists are on a rapid, inexorable march toward conquering language and meaning—as if a machine that can respond like a person in a particular context must be doing something similar to what the human brain does.

Much of what computer scientists were actually doing in this case, however, was teaching the software to identify statistical correlations in giant databases of text. For example, the terms "Massachusetts," "university," "technology," and "magazine" will often be found in documents that also contain the term "*Technology Review.*" That correlation can be calculated on the fly to answer a *Jeopardy!* question; similar methods have proved useful for search engines and automated help lines. But beyond such applications, we don't know where this particular line of research will lead, because recognizing correlations is not the same as understanding meaning; a sufficiently large statistical simulation of semantics is not the same thing as semantics. Similarly, you could use correlations and extrapolations to predict the next number in a given numeric sequence, but you need deeper analysis and mathematical proof to get it right every time.

Goodstein sequences are sequences of numbers that seem to always go up—until eventually they revert and fall to zero. A prediction based on statistical analysis of the early phase of such a sequence would get the rest of the sequence wrong. Correlations can simulate understanding without really delivering it.

Ultimately, does the Watson show really matter? Why not let IBM's PR people enjoy a day in the sun? Here's why not: there is a special danger when science is presented to the public in a sloppy way. Technical communities must exhibit exemplary behavior, because we are losing public legitimacy in the United States. Denying global climate change remains respectable in politics; many high-school biology teachers still don't fully accept evolution.

Unfortunately, the theatrics of the *Jeopardy!* contest play the same trick with neuroscience that "intelligent design" does with evolution. The facts are cast to make it seem as though they imply a metaphysical idea: in this case, that we are making machines come alive in our image.

Indeed, that is a quasi-religious idea for some technical people. There's a great deal of talk about computers inheriting the earth, perhaps in a "singularity" event—and perhaps even granting humans everlasting life in a virtual world, if we are to believe Ray Kurzweil.

But even if we quarantine overtly techno-religious ideas, the Watson-on-*Jeopardy!* scheme projects an alchemical agenda. We say, "Look, an artificial intelligence is visible in the machine's correlations." A promoter of intelligent design says, "Look, a divine intelligence is visible in the correlations derived from sources like fossils and DNA."

When we do it, how can we complain that others do it? If scientists desire respect from the public, we should expect to be emulated, and we should be careful about what methods we present for emulation.

RIVKA GALCHEN

Dream Machine

FROM THE *NEW YORKER*

*As scientists try to build quantum computers, David Deutsch, the
theorist behind them, hopes that these machines do more than just
solve complex problems with spellbinding speed. Rivka Galchen
discovers that he is betting the quantum computer will provide proof
of nothing less than the existence of multiple universes.*

O**N THE OUTSKIRTS OF** O**XFORD LIVES A BRILLIANT**
and distressingly thin physicist named David Deutsch, who
believes in multiple universes and has conceived of an as yet
unbuildable computer to test their existence. His books have titles of
colossal confidence (*The Fabric of Reality, The Beginning of Infinity*).
He rarely leaves his house. Many of his close colleagues haven't seen
him for years, except at occasional conferences via Skype.

Deutsch, who has never held a job, is essentially the founding father of quantum computing, a field that devises distinctly powerful computers based on the branch of physics known as quantum mechanics. With one millionth of the hardware of an ordinary laptop, a quantum computer could store as many bits of information as there are particles in the universe. It could break previously unbreakable codes. It could answer questions about quantum mechanics that are currently far too complicated for a regular computer to handle. None of which is to say that anyone yet knows what we would really do with one. Ask a physicist what, practically, a quantum computer would be "good for," and he might tell the story of the nineteenth-century English scientist Michael Faraday, a seminal figure in the field of electromagnetism, who, when asked how an electromagnetic effect could be useful, answered that he didn't know but that he was sure that one day it could be taxed by the Queen.

In a stairwell of Oxford's Clarendon Physics Laboratory there is a photo poster from the late nineteen-nineties commemorating the Oxford Center for Quantum Computation. The photograph shows a well-groomed crowd of physicists gathered on the lawn. Photoshopped into a far corner, with the shadows all wrong, is the head of David Deutsch, looking like a time traveller teleported in for the day. It is tempting to interpret Deutsch's representation in the photograph as a collegial joke, because of Deutsch's belief that if a quantum computer were built it would constitute near-irrefutable evidence of what is known as the Many Worlds Interpretation of quantum mechanics, a theory that proposes pretty much what one would imagine it does. A number of respected thinkers in physics besides Deutsch support the Many Worlds Interpretation, though they are a minority, and primarily educated in England, where the intense interest in quantum computing has at times been termed the Oxford flu.

But the infection of Deutsch's thinking has mutated and gone pandemic. Other scientists, although generally indifferent to the truth or falsehood of Many Worlds as a description of the universe, are now working to build these dreamed-up quantum-computing machines. Researchers at centers in Singapore, Canada, and New Haven, in collaboration with groups such as Google and NASA, may soon build machines that will make today's computers look like pocket calculators. But Deutsch complements the indifference of his colleagues to Many Worlds with one of his own—a professional indifference to the actual building of a quantum computer.

PHYSICS ADVANCES BY ACCEPTING ABSURDITIES. Its history is one of unbelievable ideas proving to be true. Aristotle quite reasonably thought that an object in motion, left alone, would eventually come to rest; Newton discovered that this wasn't true, and from there worked out the foundation of what we now call classical mechanics. Similarly, physics surprised us with the facts that the Earth revolves around the sun, time is curved, and the universe if viewed from the outside is beige.

"Our imagination is stretched to the utmost," the Nobel Prize-winning physicist Richard Feynman noted, "not, as in fiction, to imagine things which are not really there, but just to comprehend those things which are there." Physics is strange, and the people who spend their life devoted to its study are more accustomed to its strangeness than the rest of us. But, even to physicists, quantum mechanics—the basis of a quantum computer—is almost intolerably odd.

Quantum mechanics describes the natural history of matter and energy making their way through space and time. Classical mechanics does much the same, but, while classical mechanics is very accurate when describing most of what we see (sand, baseballs, planets), its descriptions of matter at a smaller scale are simply wrong. At a fine enough resolution, all those reliable rules about balls on inclined planes start to fail.

Quantum mechanics states that particles can be in two places at once, a quality called superposition; that two particles can be related, or "entangled," such that they can instantly coördinate their properties, regardless of their distance in space and time; and that when we look at particles we unavoidably alter them. Also, in quantum mechanics, the universe, at its most elemental level, is random, an idea that tends to upset people. Confess your confusion about quantum mechanics to a physicist and you will be told not to feel bad, because physicists find it confusing, too. If classical mechanics is George Eliot, quantum mechanics is Kafka.

All the oddness would be easier to tolerate if quantum mechanics merely described marginal bits of matter or energy. But it is the physics of everything. Even Einstein, who felt at ease with the idea of wormholes through time, was so bothered by the whole business that, in 1935, he co-authored a paper titled "Can quantum-mechanical description of physical reality be considered complete?" He pointed out some of quantum mechanics's strange implications, and then answered his question, essentially, in the negative. Einstein found entanglement particularly troubling, denigrating it as "spooky action at a distance," a telling phrase, which consciously echoed the seventeenth-century disparagement of gravity.

The Danish physicist Niels Bohr took issue with Einstein. He argued that, in quantum mechanics, physics had run up against the limit of what science could hope to know. What seemed like nonsense was nonsense, and we needed to realize that science, though wonderfully good at predicting the outcomes of individual experiments, could not tell us about reality itself, which would remain forever behind a veil. Science merely revealed what reality looked like to us.

Bohr's stance prevailed over Einstein's. "Of course, both sides of that dispute were wrong," Deutsch observed, "but Bohr was trying to obfuscate, whereas Einstein was actually trying to solve the problem." As Deutsch notes in *The Fabric of Reality,* "To say that prediction is the purpose of a scientific theory is to confuse means with ends. It is like saying that the purpose of a spaceship is to burn fuel." After Bohr, a "shut up and calculate" philosophy took over physics for decades. To delve into quantum mechanics as if its equations told the story of reality itself was considered sadly misguided, like those earnest inquiries people mail to 221B Baker Street, addressed to Sherlock Holmes.

I MET DAVID DEUTSCH AT his home, at four o'clock on a wintry Thursday afternoon. Deutsch grew up in the London area, took his undergraduate degree at Cambridge, stayed there for a master's in math—which he claims he's no good at—and went on to Oxford for a doctorate in physics. Though affiliated with the university, he is not on staff and has never taught a course. "I love to give talks," he told me. "I just don't like giving talks that people don't want to hear. It's wrong to set up the educational system that way. But that's not why I don't teach. I don't teach for visceral reasons—I just dislike it. If I were a biologist, I would be a theoretical biologist, because I don't like the idea of cutting up frogs. Not for moral reasons but because it's disgusting. Similarly, talking to a group of people who don't want to be there is disgusting." Instead, Deutsch has made money from lectures, grants, prizes, and his books.

In the half-light of the winter sun, Deutsch's house looked a little shabby. The yard was full of what appeared to be English ivy, and near the entrance was something twiggy and bushlike that was either dormant or dead. A handwritten sign on the door said that deliveries should "knock hard." Deutsch answered the door. "I'm very much in a rush," he told me, before I'd even stepped inside. "In a rush about so many things." His thinness contributed to an oscillation of his apparent age between nineteen and a hundred and nineteen. (He's fifty-seven.) His eyes, behind thick glasses, appeared outsized, like those of an appealing anime character. His vestibule was cluttered with old phone books, cardboard boxes, and piles of papers. "Which isn't to say that I don't have time to talk to you," he continued. "It's just that—that's why the house is in such disarray, because I'm so rushed."

More than one of Deutsch's colleagues told me about a Japanese documentary film crew that had wanted to interview Deutsch at his house. The crew asked if they could clean up the house a bit. Deutsch didn't like the idea, so the film crew promised that after filming they would reconstruct the mess as it was before. They took extensive photographs, like investigators at a crime scene, and then cleaned up. After the interview, the crew carefully reconstructed the former "disorder." Deutsch said he could still find things, which was what he had been worried about.

Taped onto the walls of Deutsch's living room were a map of the world, a periodic table, a hand-drawn cartoon of Karl Popper, a poster of the signing of the Declaration of Independence, a taxonomy of animals, a taxonomy of the characters in *The Simpsons*, color printouts of pictures of McCain and Obama, with handwritten labels reading "this one" and "that one," and two color prints of an actor who looked to me a bit like Hugh Grant. There were also old VHS tapes, an unused fireplace, a stationary exercise bike, and a large flat-screen television whose newness had no visible companion. Deutsch offered me tea and biscuits. I asked him about the Hugh Grant look-alike.

"You obviously don't watch much television," he replied. The man in the photographs was Hugh Laurie, a British actor known for his role in the American medical show *House*. Deutsch described *House* to me as "a great program about epistemology, which, apart from fundamental physics, is really my core interest. It's a program about the myriad ways that knowledge can grow or can fail to grow." Dr. House is based on Sherlock Holmes, Deutsch informed me. "And House has a friend, Wilson, who is based on Watson. Like Holmes, House is an arch-rationalist. Everything's got to have a reason, and if he doesn't know the reason it's because he doesn't know it, not because there isn't one. That's an essential attitude in fundamental science." One imagines the ghost of Bohr would disagree.

DEUTSCH'S REPUTATION AS A CLOISTERED genius stems in large part from his foundational work in quantum computing. Since the nineteen-thirties, the field of computer science has held on to the idea of a universal computer, a notion first worked out by the field's modern founder, the British polymath Alan Turing. A universal computer would be capable of comporting itself as any other computer, just as a synthesizer can make the sounds made by any other musical instrument. In a 1985 paper, Deutsch pointed out that, because Turing was working with classical physics, his universal computer could imitate only a subset of possible computers. Turing's theory needed to account for quantum mechanics if its logic was to hold. Deutsch proposed a universal computer based on quantum physics, which would have calculating powers that Turing's computer (even in theory) could not simulate.

According to Deutsch, the insight for that paper came from a conversation in the early eighties with the physicist Charles Bennett, of I.B.M., about computational-complexity theory, at the time a sexy new field that investigated the difficulty of a computational task. Deutsch questioned whether computational complexity was a fundamental or a relative property. Mass, for instance, is a fundamental property, because it remains the same in any setting; weight is a relative property, because an object's weight depends on the strength of gravity acting on it. Identical baseballs on Earth and on the moon have equivalent masses, but different weights. If computational complexity was like mass—if it was a fundamental property—then complexity was quite profound; if not, then not.

"I was just sounding off," Deutsch said. "I said they make too much of this"—meaning complexity theory—"because there's no standard computer with respect to which you should be calculating the complexity of the task." Just as an object's weight depends on the force of gravity in which it's measured, the degree of computational complexity depended on the computer on which it was measured. One could find out how complex a task was to perform on a particular computer, but that didn't say how complex a task was fundamentally, in reference to the universe. Unless there really was such a thing as a universal computer, there was no way a description of complexity could be fundamental. Complexity theorists, Deutsch reasoned, were wasting their time.

Deutsch continued, "Then Charlie said, quietly, 'Well, the thing is, there is a fundamental computer. The fundamental computer is physics itself.'" That impressed Deutsch. Computational complexity was a fundamental property; its value referenced how complicated a computation was on that most universal computer, that of the physics of the world. "I realized that Charlie was right about that," Deutsch said. "Then I thought, But these guys are using the wrong physics. They realized that complexity theory was a statement about physics, but they didn't realize that it mattered whether you used the true laws of physics, or some approximation, i.e., classical physics." Deutsch began rewriting Turing's universal-computer work using quantum physics. "Some of the differences are very large," he said. Thus, at least in Deutsch's mind, the quantum universal computer was born.

A number of physics journals rejected some of Deutsch's early quantum-computing work, saying it was "too philosophical." When it was finally published, he said, "a handful of people kind of got it." One of them was the physicist Artur Ekert, who had come to Oxford as a graduate student, and who told me, "David was really the first one who formulated the concept of a quantum computer."

Other important figures early in the field included the reclusive physicist Stephen J. Wiesner, who, with Bennett's encouragement, developed ideas like quantum money (uncounterfeitable!) and quantum cryptography, and the philosopher of physics David Albert, whose imagining of introspective quantum automata (think robots in analysis) Deutsch describes in his 1985 paper as an example of "a true quantum computer." Ekert says of the field, "We're a bunch of odd ducks."

Although Deutsch was not formally Ekert's adviser, Ekert studied with him. "He kind of adopted me," Ekert recalled, "and then, afterwards, I kind of adopted him. My tutorials at his place would start at around 8 P.M., when David would be having his lunch. We'd stay talking and working until the wee hours of the morning. He likes just talking things over. I would leave at 3 or 4 A.M., and then David would start properly working afterwards. If we came up with something, we would write the paper, but sometimes we wouldn't write the paper, and if someone else also came up with the solution we'd say, 'Good, now we don't have to write it up.'" It was not yet clear, even in theory, what a quantum computer might be better at than a classical computer, and so Deutsch and Ekert tried to develop algorithms for problems that were intractable on a classical computer but that might be tractable on a quantum one.

One such problem is prime factorization. A holy grail of mathematics for centuries, it is the basis of much current cryptography. It's easy to take two large prime numbers and multiply them, but it's very difficult to take a large number that is the product of two primes and then deduce what the original prime factors are. To factor a number of two hundred digits or more would take a regular computer many lifetimes. Prime factorization is an example of a process that is easy one way (easy to scramble eggs) and very difficult the other (nearly impossible to unscramble them). In cryptography, two large prime numbers are multiplied to create a security key. Unlocking that key would be the equivalent of unscrambling an egg. Using prime factorization in this way is called RSA encryption (named for the scientists who proposed it, Rivest, Shamir, and Adleman), and it's how most everything is kept secret on the Internet, from your credit-card information to I.R.S. records.

In 1992, the M.I.T. mathematician Peter Shor heard a talk about theoretical quantum computing, which brought to his attention the work of Deutsch and other foundational thinkers in what was then still an obscure field. Shor worked on the factorization problem in private. "I wasn't sure anything would come of it," Shor explained. But, about a year later, he emerged with an algorithm that (a) could only be run on a quantum computer, and (b) could quickly find the prime factors of a very large number—the grail! With Shor's algorithm, calculations that would take a normal computer longer than the history of the universe would take a sufficiently powerful quantum computer an afternoon. "Shor's work was the biggest jump," the physicist David DiVincenzo, who is considered among the most knowledgeable about the history of quantum computing, says. "It was the moment when we were, like, Oh, now we see what it would be good for."

Today, quantum computation has the sustained attention of experimentalists; it also has serious public and private funding. Venture-capital companies are already investing in quantum encryption devices, and university research groups around the world have large teams working both to build hardware and to develop quantum-computer applications—for example, to model proteins, or to better understand the properties of superconductors.

Artur Ekert became a key figure in the transition from pure theory to building machines. He founded the quantum computation center at Oxford, as well as a similar center a few years later at Cambridge. He now leads a center in Singapore, where the government has made quantum-computing research one of its top goals. "Today in the field there's a lot of focus on lab implementation, on how and from what you could actually build a quantum computer," DiVincenzo said. "From the perspective of just counting, you can say that the majority of the field now is involved in trying to build some hardware. That's a result of the success of the field." In 2009, Google announced that it had been working on quantum-computing algorithms for three years, with the aim of having a computer that could quickly identify particular things or people from among vast stores of video and images—David Deutsch, say, from among millions of untagged photographs.

IN THE EARLY NINETEENTH CENTURY, a "computer" was any person who computed: someone who did the math for building a bridge, for example. Around 1830, the English mathematician and inventor Charles Babbage worked out his idea for an Analytical Engine, a machine that would remove the human from computing, and thus bypass human error. Nearly no one imagined an analytical engine would be of much use, and in Babbage's time no such machine was ever built to completion. Though Babbage was prone to serious mental breakdowns, and though his bent of mind was so odd that he once wrote to Alfred Lord Tennyson correcting his math (Babbage suggested rewriting "Every minute dies a man / Every minute one is born" as "Every moment dies a man / Every moment one and a sixteenth is born," further noting that although the exact figure was 1.167, "something must, of course, be conceded to the laws of meter")—we can now say the guy was on to something.

A classical computer—any computer we know today—transforms an input into an output through nothing more than the manipulation of binary bits, units of information that can be either zero or one. A quantum computer is in many ways like a regular computer, but instead of bits it uses qubits. Each qubit (pronounced "Q-bit") can be zero or one, like a bit, but a qubit can also be zero *and* one— the quantum-mechanical quirk known as superposition. It is the state that the cat in the classic example of Schrödinger's closed box is stuck in: dead and alive at the same time. If one reads quantum-mechanical equations literally, superposition is ontological, not epistemological; it's not that we don't know which state the cat is in, but that the cat really is in both states at once. Superposition is like Freud's description of true ambivalence: not feeling unsure, but feeling opposing extremes of conviction at once. And, just as ambivalence holds more information than any single emotion, a qubit holds more information than a bit.

What quantum mechanics calls entanglement also contributes to the singular powers of qubits. Entangled particles have a kind of E.S.P.: regardless of distance, they can instantly share information that an observer cannot even perceive is there. Input into a quantum computer can thus be dispersed among entangled qubits, which lets the processing of that information be spread out as well: tell one particle something, and it can instantly spread the word among all the other particles with which it's entangled.

There's information that we can't perceive when it's held among entangled particles; that information is their collective secret. As quantum mechanics has taught us, things are inexorably changed by our trying to ascertain anything about them. Once observed, qubits are no longer in a state of entanglement, or of superposition: the cat commits irrevocably to life or death, and this ruins the quantum computer's distinct calculating power. A quantum computer is the pot that, if watched, really won't boil. Charles Bennett described quantum information as being "like the information of a dream—we can't show it to others, and when we try to describe it we change the memory of it."

But, once the work on the problem has been done among the entangled particles, then we can look. When one turns to a quantum computer for an "answer," that answer, from having been held in that strange entangled way, among many particles, needs then to surface in just one, ordinary, unentangled place. That transition from entanglement to non-entanglement is sometimes termed "collapse." Once the system has collapsed, the information it holds is no longer a dream or a secret or a strange cat at once alive and dead; the answer is then just an ordinary thing we can read off a screen.

QUBITS ARE NOT MERELY THEORETICAL. Early work in quantum-computer hardware built qubits by manipulating the magnetic nuclei of atoms in a liquid soup with electrical impulses. Later teams, such as the one at Oxford, developed qubits using single trapped ions, a method that confines charged atomic particles to a particular space. These qubits are very precise, though delicate; protecting them from interference is quite difficult. More easily manipulated, albeit less precise, qubits have been built from super-conducting materials arranged to model an atom. Typically, the fabrication of a qubit is not all that different from that of a regular chip. At Oxford, I saw something that resembled an oversize air-hockey table chaotically populated with a specialty Lego set, with what looked like a salad-bar sneeze guard hovering over it; this extended apparatus comprised lasers and magnetic-field generators and optical cavities, all arranged at just the right angles to manipulate and protect from interference the eight tiny qubits housed in a steel tube at the table's center.

Oxford's eight-qubit quantum computer has significantly less computational power than an abacus, but fifty to a hundred qubits could make something as powerful as any laptop. A team in Bristol, England, has a small, four-qubit quantum computer that can factor the number 15. A Canadian company claims to have built one that can do Sudoku, though that has been questioned by some who say that the processing is effectively being done by normal bits, without any superposition or entanglement.

Increasing the number of qubits, and thus the computer's power, is more than a simple matter of stacking. "One of the main problems with scaling up is a qubit's fidelity," Robert Schoelkopf, a physics professor at Yale who leads a quantum-computing team, explained. By fidelity, he refers to the fact that qubits "decohere"—fall out of their information-holding state—very easily. "Right now, qubits can be faithful for about a microsecond. And our calculations take about one hundred nanoseconds. Either calculations need to go faster or qubits need to be made more faithful."

WHAT QUBITS ARE DOING AS we avert our gaze is a matter of some dispute, and occasionally—"shut up and calculate"—of some determined indifference, especially for more pragmatically minded physicists. For Deutsch, to really understand the workings of a quantum computer necessitates subscribing to Hugh Everett's Many Worlds Interpretation of quantum mechanics. Everett's theory was neglected upon its publication, in 1957, and is still a minority view. It entails the following counterintuitive reasoning: every time there is more than one possible outcome, all of them occur. So if a radioactive atom might or might not decay at any given second, it both does and doesn't; in one universe it does, and in another it doesn't. These small branchings of possibility then ripple out until everything that is possible in fact is. According to Many Worlds theory, instead of a single history there are innumerable branchings. In one universe your cat has died, in another he hasn't, in a third you died in a sledding accident at age seven and never put your cat in the box in the first place, and so on.

Many Worlds is an ontologically extravagant proposition. But it also bears some comfortingly prosaic implications: in Many Worlds theory, science's aspiration to explain the world fully remains intact. The strangeness of superposition is, as Deutsch explains it, simply "the phenomenon of physical variables having different values in different universes." And entanglement, which so bothered Einstein and others, especially for its implication that particles could instantly communicate regardless of their distance in space or time, is also resolved. Information that seemed to travel faster than the speed of light and along no detectable pathway—spookily transmitted as if via E.S.P.—can, in Many Worlds theory, be understood to move differently. Information still spreads through direct contact—the "ordinary" way; it's just that we need to adjust to that contact being via the tangencies of abutting universes. As a further bonus, in Many Worlds theory randomness goes away, too. A ten-per-cent chance of an atom decaying is not arbitrary at all, but rather refers to the certainty that the atom will decay in ten per cent of the universes branched from that point. (This being science, there's the glory of nuanced dissent around the precise meaning of each descriptive term, from "chance" to "branching" to "universe.")

In the nineteen-seventies, Everett's theory received some of the serious attention it missed at its conception, but today the majority of physicists are not much compelled. "I've never myself subscribed to that view," DiVincenzo says, "but it's not a harmful view." Another quantum-computing physicist called it "completely ridiculous," but Ekert said, "Of all the weird theories out there, I would say Many Worlds is the least weird." In Deutsch's view, "Everett's approach was to look at quantum theory and see what it actually said, rather than hope it said certain things. What we want is for a theory to conform to reality, and, in order to find out whether it does, you need to see what the theory actually says. Which with the deepest theories is actually quite difficult, because they violate our intuitions."

I told Deutsch that I'd heard that even Everett thought his theory could never be tested.

"That was a catastrophic mistake," Deutsch said. "Every innovator starts out with the world view of the subject as it was before his innovation. So he can't be blamed for regarding his theory as an interpretation. But"—and here he paused for a moment—"I proposed a test of the Everett theory."

Deutsch posited an artificial-intelligence program run on a computer which could be used in a quantum-mechanics experiment as an "observer"; the A.I. program, rather than a scientist, would be doing the problematic "looking," and, by means of a clever idea that Deutsch came up with, a physicist looking at the A.I. observer would see one result if Everett's theory was right, and another if the theory was wrong.

It was a thought experiment, though. No A.I. program existed that was anywhere near sophisticated enough to act as the observer.

Deutsch argued that theoretically there could be such a program, though it could only be run on radically more advanced hardware— hardware that could model any other hardware, including that of the human brain. The computer on which the A.I. program would run "had to have the property of being universal . . . so I had to postulate this quantum-coherent universal computer, and that was really my first proposal for a quantum computer. Though I didn't think of it as that. And I didn't call it a quantum computer. But that's what it was." Deutsch had, it seems, come up with the idea for a quantum computer twice: once in devising a way to test the validity of the Many Worlds Interpretation, and a second time, emerging from the complexity-theory conversation, with evidenced argument supporting Many Worlds as a consequence.

TO THOSE WHO FIND THE Many Worlds Interpretation needlessly baroque, Deutsch writes, "the quantum theory of parallel universes is not the problem—it is the solution. . . . It is the explanation—the only one that is tenable—of a remarkable and counterintuitive reality." The theory also explains how quantum computers might work. Deutsch told me that a quantum computer would be "the first technology that allows useful tasks to be performed in collaboration between parallel universes." The quantum computer's processing power would come from a kind of outsourcing of work, in which calculations literally take place in other universes. Entangled particles would function as paths of communication among different universes, sharing information and gathering the results. So, for example, with the case of Shor's algorithm, Deutsch said, "When we run such an algorithm, countless instances of us are also running it in other universes. The computer then differentiates some of those universes (by creating a superposition) and as a result they perform part of the computation on a huge variety of different inputs. Later, those values affect each other, and thereby all contribute to the final answer, in just such a way that the same answer appears in all the universes."

Deutsch is mainly interested in the building of a quantum computer for its implications for fundamental physics, including the Many Worlds Interpretation, which would be a victory for the argument that science can explain the world and that, consequently, reality is knowable. ("House cures people," Deutsch said to me when discussing Hugh Laurie, "because he's interested in solving problems, not because he's interested in people.") Shor's algorithm excites Deutsch, but here is how his excitement comes through in his book *The Fabric of Reality*:

> To those who still cling to a single-universe world-view, I issue this challenge: *explain how Shor's algorithm works.* I do not merely mean predict that it will work, which is merely a matter of solving a few uncontroversial equations. I mean provide an explanation. When Shor's algorithm has factorized a number, using 10^{500} or so times the computational resources than can be seen to be present, where was the number factorized? There are only about 10^{80} atoms in the entire visible universe, an utterly minuscule number compared with 10^{500}. So if the visible universe were the extent of physical reality, physical reality would not even remotely contain the resources required to factorize such a large number. Who did factorize it, then? How, and where, was the computation performed?

Deutsch believes that quantum computing and Many Worlds are inextricably bound. He is nearly alone in this conviction, though many (especially around Oxford) concede that the construction of a sizable and stable quantum computer might be evidence in favor of the Everett interpretation. "Once there are actual quantum computers," Deutsch said to me, "and a journalist can go to the actual labs and ask how does that actual machine work, the physicists in question will then either talk some obfuscatory nonsense, or will explain it in terms of parallel universes. Which will be newsworthy. Many Worlds will then become part of our culture. Really, it has nothing to do with making the computers. But psychologically it has everything to do with making them."

It's tempting to view Deutsch as a visionary in his devotion to the Many Worlds Interpretation, for the simple reason that he has been a visionary before. "Quantum computers should have been invented in the nineteen-thirties," he observed near the end of our conversation. "The stuff that I did in the late nineteen-seventies and early nineteen-eighties didn't use any innovation that hadn't been known in the thirties." That is straightforwardly true. Deutsch went on, "The question is why."

DiVincenzo offered a possible explanation. "Your average physicists will say, 'I'm not strong in philosophy and I don't really know what to think, and it doesn't matter.'" He does not subscribe to Many Worlds, but is reluctant to dismiss Deutsch's belief in it, partly because it has led Deutsch to come up with his important theories, but also because "quantum mechanics does have a unique place in physics, in that it does have a subcurrent of philosophy you don't find even in Newton's laws or gravity. But the majority of physicists say it's a quagmire they don't want to get into—they'd rather work out the implications of ideas; they'd rather calculate something."

AT YALE, A TEAM LED by Robert Schoelkopf has built a two-qubit quantum computer. "Deutsch is an original thinker and those early papers remain very important," Schoelkopf told me. "But what we're doing here is trying to develop hardware, to see if these descriptions that theorists have come up with work." They have configured their computer to run what is known as a Grover's algorithm, one that deals with a four-card-monte type of question: Which hidden card is the queen? It's a sort of Shor's algorithm for beginners, something that a small quantum computer can take on.

The Yale team fabricates their qubit processor chips in house.

"The chip is basically made of a very thin wafer of sapphire or silicon—something that's a good insulator—that we then lay a patterned film of superconducting metal on to form the wiring and qubits," Schoelkopf said. What they showed me was smaller than a pinkie nail and looked like a map of a subway system.

Schoelkopf and his colleague Michel Devoret, who leads a separate team, took me to a large room of black lab benches, inscrutable equipment, and not particularly fancy monitors. The aesthetic was inadvertent steampunk. The dust in the room made me sneeze. "We don't like the janitors to come sweep for fear they'll disturb something," Schoelkopf said.

The qubit chip is small, but its supporting apparatus is imposing. The largest piece of equipment is the plumbing of the very high-end refrigerator, which reduces the temperature around the two qubits to ten millidegrees above absolute zero. The cold improves the computer's fidelity. Another apparatus produces the microwave signals that manipulate the qubits and set them into any degree of superposition that an experimenter chooses.

Running this Grover's algorithm takes a regular computer three or fewer steps—if after checking the third card you still haven't found the queen, you know she is under the fourth card—and on average it takes 2.25 steps. A quantum computer can run it in just one step. This is because the qubits can represent different values at the same time. In the four-card-monte example, each of the cards is represented by one of four states: 0,0; 0,1; 1,0; 1,1. Schoelkopf designates one of these states as the queen, and the quantum computer must determine which one. "The magic comes from the initial state of the computer," he explained. Both of the qubits are set up, via pulses of microwave radiation, in a superposition of zero and one, so that each qubit represents two states at once, and together the two qubits represent all four states.

"Information can, in a way, be holographically represented across the whole computer; that's what we exploit," Devoret explained.

"This is a property you don't find in a classical information processor. A bit has to be in one state—it has to be here or there. It's useful to have the bit be everywhere."

Through superposition and entanglement, the computer simultaneously investigates each of the four possible queen locations. "Right now we only get the right answer eighty per cent of the time, and we find even that pretty exciting," Schoelkopf said.

With Grover's algorithm, or theoretically with Shor's, calculations are performed in parallel, though not necessarily in parallel worlds. "It's as if I had a gazillion classical computers that were all testing different prime factors at the same time," Schoelkopf summarized. "You start with a well-defined state, and you end with a well-defined state. In between, it's a crazy entangled state, but that's fine."

Schoelkopf emphasized that quantum mechanics is a funny system but that it really is correct. "These oddnesses, like superposition and entanglement—they seemed like limitations, but in fact they are exploitable resources. Quantum mechanics is no longer a new or surprising theory that should strike us as odd."

Schoelkopf seemed to suggest that existential questions like those which Many Worlds poses might be, finally, simply impracticable. "If you have to describe a result in my lab in terms of the computing chip," he continued, "plus the measuring apparatus, plus the computer doing data collection, plus the experimenter at the bench . . . at some point you just have to give up and say, Now quantum mechanics doesn't matter anymore, now I just need a classical result. At some point you have to simplify, you have to throw out some of the quantum information." When I asked him what he thought of Many Worlds and of "collapse" interpretations—in which "looking" provokes a shift from an entangled to an unentangled state—he said, "I have an alternate language which I prefer in describing quantum mechanics, which is that it should really be called Collapse of the Physicist." He knows it's a charming formulation, but he does mean something substantive in saying it. "In reality it's about where to collapse the discussion of the problem."

I thought Deutsch might be excited by the Yale team's research, and I e-mailed him about the progress in building quantum computers. "Oh, I'm sure they'll be useful in all sorts of ways," he replied. "I'm really just a spectator, though, in experimental physics."

SIR ARTHUR CONAN DOYLE NEVER liked detective stories that built their drama by deploying clues over time. Conan Doyle wanted to write stories in which all the ingredients for solving the crime were there from the beginning, and in which the drama would be, as in the Poe stories that he cited as precedents, in the mental workings of his ideal ratiocinator. The story of quantum computing follows a Holmesian arc, since all the clues for devising a quantum computer have been there essentially since the discovery of quantum mechanics, waiting for a mind to properly decode them.

But writers of detective stories have not always been able to hew to the rationality of their idealized creations. Conan Doyle believed in "spiritualism" and in fairies, even as the most famed spiritualists and fairy photographers kept revealing themselves to be fakes. Conan Doyle was also convinced that his friend Harry Houdini had supernatural powers; Houdini could do nothing to persuade him otherwise. Conan Doyle just knew that there was a spirit world out there, and he spent the last decades of his life corralling evidence ex post facto to support his unshakable belief.

Physicists are ontological detectives. We think of scientists as wholly rational, open to all possible arguments. But to begin with a conviction and then to use one's intellectual prowess to establish support for that conviction is a methodology that really has worked for scientists, including Deutsch. One could argue that he dreamed up quantum computing because he was devoted to the idea that science can explain the world. Deutsch would disagree.

In *The Fabric of Reality*, Deutsch writes, "I remember being told, when I was a small child, that in ancient times it was still possible to know everything that was known. I was also told that nowadays so much is known that no one could conceivably learn more than a tiny fraction of it, even in a long lifetime. The latter proposition surprised and disappointed me. In fact, I refused to believe it." Deutsch's life's work has been an attempt to support that intuitive disbelief—a gathering of argument for a conviction he held because he just knew.

Deutsch is adept at dodging questions about where he gets his ideas. He joked to me that they came from going to parties, though I had the sense that it had been years since he'd been to one. He said, "I don't like the style of science reporting that goes over that kind of thing. It's misleading. So Brahms lived on black coffee and forced himself to write a certain number of lines of music a day. Look," he went on, "I can't stop you from writing an article about a weird English guy who thinks there are parallel universes. But I think that style of thinking is kind of a put-down to the reader. It's almost like saying, If you're not weird in these ways, you've got no hope as a creative thinker. That's not true. The weirdness is only superficial."

Talking to Deutsch can feel like a case study of reason following desire; the desire is to be a creature of pure reason. As he said in praise of Freud, "He did a good service to the world. He made it O.K. to speak about the mechanisms of the mind, some of which we may not be aware of. His actual theory was all false, there's hardly a single true thing he said, but that's not so bad. He was a pioneer, one of the first who tried to think about things rationally."

Linda Marsa

Going to Extremes

FROM *DISCOVER*

*Australia's extreme climate has intensified in recent years—
catastrophic flooding and storms with winds up to 180 miles per
hour have brought record damage. Linda Marsa asks whether this is
what global warming has in store for the rest of the world.*

T HE RIVER CAME UP TO RIGHT WHERE WE'RE SITTING,
and the waters were more than two feet deep," Peter Goodwin tells me in the driveway of his ranch-style house perched
on the banks of the Balonne River in St. George, a village of 3,500 in
eastern Australia. It is a drizzly Sunday afternoon in April, three
months after a devastating flood that drenched a landmass the size
of France and Germany combined and isolated the town after the
rain-swollen river rose to a record 45 feet.

Agricultural areas like St. George were hardest hit by the relentless rains and overflowing rivers that swamped roads, cut off power lines, washed away vineyards and fruit orchards, drowned thousands of head of cattle and other livestock, and covered homes and everything inside them in thick layers of sediment and mud. Shell-shocked residents are still digging out from under the debris.

"That's the hard part of the flood—the aftermath," says Goodwin, 60, a crusty, compactly built man with piercing blue eyes and calloused hands who works as an operations manager for the local municipality and has been staying with his grown daughter while he makes his home habitable again. "You get a lot of help during the flood, but then everyone settles back into their routine. There are a lot of houses down there that are still empty," he adds, gesturing toward the riverbank. "And they will be for a long time to come."

Everywhere I travel down under, the stoic Aussies are industriously patching together their shattered country, which has been walloped by one weather-related disaster after another. On an overcast Monday morning, I'm 200 miles east of St. George, cruising toward Brisbane, a sophisticated coastal city of more than 2 million, when I hit traffic that's backed up for miles on the two-lane blacktop that crosses Cunningham's Gap. Construction crews are still repairing the flood-related destruction on this highway and other roads throughout much of eastern Australia. A line of passenger cars, along with the brawny pickups favored in rural areas and the aptly named "road trains"—the double-trailer 16-wheelers that ferry cotton, apples, grapes, peaches, pears, and other produce to the coastal cities—inch down the steep grade, stuck in the kind of slow-motion gridlock that would normally send tempers flaring and horns blaring. After almost an hour in that slowdown, steam is coming out of my ears, but the other drivers seem to accept the jammed roads with a remarkable equanimity, looking upon them as simply another step in the process of recovery.

Australia has always been a country of climate extremes, but lately the swings of wet and dry have been shockingly intense. Weeks of torrential rains at the end of 2010 and beginning of 2011 (summertime down under) created tsunami-like waters that covered most of the province of Queensland. Then in mid-January, a cloudburst from a freak storm caused the Brisbane River to overflow its banks. A surging wall of water gushed through the pocket parks and tree-lined streets of downtown Brisbane, saturating trendy boutiques, stately government buildings, and gleaming glass skyscrapers.

As the waters advanced, thousands of people evacuated their homes and businesses. Panicked shoppers emptied supermarket shelves. With most roads closed, fleeing residents jammed an open highway in a frantic exodus reminiscent of the chaos on arteries leading out of New Orleans when the twin disasters of Katrina and the breached levees filled that city like a bathtub. Choppers crowded the skies ferrying supplies and rescuing stranded residents off rooftops. Lines for sandbags stretched outside relief centers, which were filled to capacity with people left homeless, and troops trucked in fresh food, water, and emergency supplies to a city the size of Houston.

The flooding claimed 35 lives and left more than 100,000 people homeless. The waters destroyed food crops, pushing prices of staples like bananas to $7 (AUD) a pound and decimating vineyards. Even the country's lucrative coal mining industry was brought to a standstill. Damage came to an estimated $20 billion overall.

Less than three weeks later, nature struck again. In February, Cyclone Yasi, a Category 5 storm with a 300-mile front and winds gusting up to 180 miles an hour, roared in from the Coral Sea and slammed into Australia's northeastern coast, flattening several beachfront towns still cleaning up from the floods. The largest cyclone to hit Queensland in nearly a century, Yasi forced thousands of people to flee Cairns, a steamy tropical paradise of 160,000 (think Miami Beach in the 1950s, but without the movie stars). Parts of the city were without power for five days; hospitals were emptied and patients were airlifted 1,000 miles south to Brisbane. Many villages along the coast, as well as the banana and sugar plantations, were left in shambles. From the air, the coastline looked as if a pack of 3-year-olds had gone on a rampage with Tinker Toys.

AUSTRALIA IS OFTEN CALLED THE lucky country. It is flush with natural resources—uranium, coal, oil, gold, and the rare earth minerals that are used in cell phones and electronics—and blessed with sparkling, pristine beaches that extend thousands of miles. It is home to some of the world's oldest rainforests, lush tropical jungles with dense canopies that cover the northeastern tip of the continent. It is also a spacious country, almost as large as the continental United States, but with a population of just 22 million, smaller than that of Texas. About half of those people are clustered in three major cities that hug the eastern coast: Sydney, Melbourne, and Brisbane.

But the floods of the past year hint at a new, less fortunate chapter in Australia's history. On many fronts, Oz is a land under siege. A changing climate seems to be intensifying the long-standing cycle of drought and flood—a state of affairs that could magnify the gap between the continent's soggy coasts and arid interior in the future. In addition to the flooding, recent years have brought record heat waves, costly crop losses, and brush fires of unimaginable ferocity. As the planet heats up, these events are expected to become more frequent and severe.

Australia exemplifies what global warming models have long predicted, climate experts say, and what is happening in this hardy nation offers a glimpse of some of what is on the horizon for the United States as well. Rising temperatures tend to produce more extreme weather events, according to the models. "Our recent extremes may be strongly climate change-driven," says Penny Whetton, a climatologist with the Commonwealth Scientific and

Industrial Research Organization (CSIRO), Australia's national science agency, in Victoria. "They're an illustration of what we expect to see more of in the future, when natural fluctuations are intensified by global warming."

WITH WARMER TEMPERATURES WILL COME more floods like the ones that have buffeted Australia over the past year, but the other half of the climate equation is making itself abundantly felt as well. During the past decade, much of Australia was gripped by a fierce dry spell that sparked stringent water rationing measures throughout the southern and eastern parts of the country. Evening news shows regularly announced water levels in the nation's reservoirs and featured stories about scofflaws being slapped with heavy fines after neighbors caught them hosing down their cars. In one celebrated incident, South Australia's water minister quenched his thirsty lawn in flagrant defiance of the rules, triggering a public outcry.

Adelaide, the driest major city in the world's driest inhabited continent, came close to running out of drinking water in 2009, and officials were days away from importing bottled water for some of the area's million-plus residents. Farmlands in the once fertile Murray-Darling Basin—a region in Southeastern Australia the size of France and Spain combined—were turned into cracked dust bowls. More than 10,000 families, many of whom had farmed for generations, were forced off the land. Power plants were shut down for lack of cooling water. Record heat waves baked the soil, killed wildlife, and turned the bone-dry terrain into kindling for firestorms that incinerated entire towns.

Australia prominently wears the scars of these escalating skirmishes with nature. The nation's capital, Canberra, is a parklike planned city of nearly 360,000 that occupies a patch of land partway between Sydney and Melbourne, with wide, tree-lined boulevards that radiate outward from Lake Burley Griffin. Yet today, an aerial view reveals huge swaths of the countryside surrounding Canberra blackened from a firestorm that threatened to engulf the city in 2003, after months of scorching weather turned pine forests into a giant tinderbox.

About 20 miles outside Canberra is the Googong Reservoir, a long finger of water that stretches more than 650 acres when filled to the brim. More than a million people get their drinking water from the reservoir, but the drought shriveled it to less than 30 percent of its capacity. Even after the recent drenching rains, water levels remain perilously low.

Elizabeth Hanna, a public health researcher at Australia National University in Canberra, has a hard time reconciling another shrunken body of water, Lake George, with the shimmering sea of her youth. "When I was a child, we'd drive up from Melbourne and it would be overflowing its banks," she recalls.

SOME OF AUSTRALIA'S SENSITIVITY TO climate change stems from its evolution as a continent. Around 45 million years ago, the Australian landmass broke away from Gondwana, an ancient super-continent that encompassed South America, Africa, and Antarctica as well. Almost completely covered in rainforest in those days, Australia then drifted northward toward the equator in complete isolation for 30 million years. During that period, the central regions of Australia dried out and created some of the world's most ancient deserts.

Australian soils are deficient in vital nutrients including nitrogen, phosphorus, and zinc, mainly because the region is so old, even in geologic time, and most of the country has not been revitalized by the soil-renewing activity of volcanoes or glaciers for tens of millions of years. "The country is already dry and ecologically fragile, with relatively poor soils," says Steven Sherwood, an atmospheric physicist and codirector of the Climate Change Research Centre at the University of New South Wales in Sydney. "It doesn't have the resilience, so it's more vulnerable to anything."

Modern development has exacerbated some of those vulnerabilities. River systems have been heavily exploited for irrigation, which has hastened soil degradation, made the ground saltier, and contributed to creeping desertification. And roughly 85 percent of the population is crammed into coastal cities that are susceptible to storm surges and cyclones. But the prime driver of Australia's unique vulnerability is geographic: its location in the midst of three oceans, and the interacting currents that result.

Pacific currents have historically switched between patterns known as El Niño and La Niña, triggering correspondingly dry and wet years and perhaps decades. During El Niño, warmer waters in the southeastern Pacific create air pressure systems that drive rain out over the ocean, resulting in less precipitation over the continent itself. La Niña cools ocean temperatures, leading to stronger easterly winds in the tropics. All this causes more water to evaporate into the atmosphere, intensifying monsoon rains and cyclones. At the same time, the Indian Ocean Dipole, a current discovered only in the past decade, cools the eastern tropics and warms western currents, lessening spring rains in the southeast.

Rising global temperatures portend shifts in all these ocean currents, potentially with drastic consequences, says Albert Gabric, an environmental scientist at Griffith University in Brisbane. Some of the strongest El Niño patterns ever recorded were blamed for the crippling drought, while an equally intense La Niña brought on the downpours that inundated Queensland. Changes in the Southern Annular Mode, an ocean current that prevents low-pressure rain systems from passing over southern Australia, has helped shut off winter showers in Perth. Rainfall over this area has dropped by 20 percent.

Beyond these perturbations, there are worrisome hints that water is already tunneling away underneath the ice shelf in Antarctica.

"Breaking up of glaciers in Antarctica will cool the surface temperature of the Southern Ocean, which in turn could make wind patterns more intense, causing more severe and frequent cyclones," Gabric says. "There are multiple stressors, creating layers upon layers of complexity, and warming adds an extra dimension."

To an American visitor, Australia feels like a parallel universe. Here in the United States, politicians and radio talk-show hosts regularly ridicule the notion of global warming and attack the integrity of climate scientists. Aussies, in stark contrast, have little doubt that they face increasingly stormy weather and are bracing for a bumpy ride. "You can't live in this area and not realize what is happening to the land," says Steve Turton, a physical geographer at James Cook University in Cairns.

In many ways, Australia is still grappling with how to respond to climate change. It is a coal superpower, the world's largest exporter of the dirtiest of all fossil fuels. Australia relies heavily on coal for its own electricity as well, emitting more CO_2 per person than any other developed country, and its agricultural emissions are among the highest per capita in the world, mainly because of the large numbers of sheep and cattle.

But a recently introduced carbon tax—a possible model for other industrialized nations—creates monetary incentives for curbing the use of fossil fuels and investing more in renewable energy. The Australian government recently invested nearly $500 million in climate change research, including $126 million specifically earmarked to identify the best strategies for adaptation to hotter days, storm surges, drought, and rising sea levels. Scientists are actively working on drought-resistant crops and farming practices, better ways to use and distribute drinking water, new firefighting strategies, and techniques for preserving Australia's unique ecosystems and wildlife at an affordable price.

"We are between a rock and a hard place," says Christopher Cocklin, Senior Deputy Vice Chancellor of James Cook University in Townsville. "We have an economy that is deeply reliant on fossil fuel resources such as coal, but we are getting religion. With the right amount of will and leadership we can certainly avoid the worst."

Australians did take some convincing, though, especially since severe weather swings were already part of the country's daily life. An emphatic 2008 report by economist Ross Garnaut, a former global warming agnostic who became, in his own words, "a late-life convert" to the green cause, did much to dispel any lingering questions among most Australians about whether the threat of climate change was real.

"There is a resistance in the emissions-intensive [industries] to change," Garnaut said in a keynote speech at the Greenhouse 2011 conference held earlier this year in Cairns. "I understand the preference to wish it away. But if we do nothing . . . it will mean the end of civilization as we know it. We'll be one of the species facing extinction."

JEFF GOODELL

The Fire Next Time

FROM *ROLLING STONE*

> *In the wake of the Fukushima nuclear disaster, Jeff Goodell investi-*
> *gates the safety of America's aging collection of nuclear power*
> *plants—and finds that what happened in Japan could happen*
> *here.*

FIVE DAYS AFTER A MASSIVE EARTHQUAKE AND TSU-
nami struck Japan, triggering the worst nuclear disaster since
Chernobyl, America's leading nuclear regulator came before
Congress bearing good news: Don't worry, it can't happen here. In
the aftermath of the Japanese catastrophe, officials in Germany
moved swiftly to shut down old plants for inspection, and China put
licensing of new plants on hold. But Gregory Jaczko, the chairman of
the Nuclear Regulatory Commission, reassured lawmakers that

nothing at the Fukushima Daiichi reactors warranted any immediate changes at U.S. nuclear plants. Indeed, 10 days after the earthquake in Japan, the NRC extended the license of the 40-year-old Vermont Yankee nuclear reactor—a virtual twin of Fukushima—for another two decades. The license renewal was granted even though the reactor's cooling tower had literally fallen down, and the plant had repeatedly leaked radioactive fluid.

Perhaps Jaczko was simply trying to prevent a full-scale panic about the dangers of U.S. nuclear plants. After all, there are now 104 reactors scattered across the country, generating 20 percent of America's power. All of them were designed in the 1960s and '70s, and are nearing the end of their planned life expectancy. But there was one problem with Jaczko's testimony, according to Dave Lochbaum, a senior adviser at the Union of Concerned Scientists: Key elements of what the NRC chief told Congress were "a baldfaced lie."

Lochbaum, a nuclear engineer, says that Jaczko knows full well that what the NRC calls "defense in depth" at U.S. reactors has been seriously compromised over the years. In some places, highly radioactive spent fuel is stockpiled in what amounts to swimming pools located beside reactors. In other places, changes in the cooling systems at reactors have made them more vulnerable to a core meltdown if something goes wrong. A few weeks before Fukushima, Lochbaum authored a widely circulated report that underscored the NRC's haphazard performance, describing 14 serious "near-miss" events at nuclear plants last year alone. At the Indian Point reactor just north of New York City, federal inspectors discovered a water-containment system that had been leaking for 16 years.

As head of the NRC, Jaczko is the top cop on the nuclear beat, the guy charged with keeping the nation's fleet of aging nukes running safely. A balding, 40-year-old Democrat with big ears and the air of a brilliant high school physics teacher, Jaczko oversees a 4,000-person agency with a budget of $1 billion. But the NRC has long served as little more than a lap dog to the nuclear industry, unwilling to crack down on unsafe reactors. "The agency is a wholly owned subsidiary of the nuclear power industry," says Victor Gilinsky, who served on the commission during the Three Mile Island meltdown in 1979. Even President Obama denounced the NRC during the 2008 campaign, calling it a "moribund agency that needs to be revamped and has become captive of the industries that it regulates."

In the years ahead, nuclear experts warn, the consequences of the agency's inaction could be dire. "The NRC has consistently put industry profits above public safety," says Arnie Gundersen, a former nuclear executive turned whistle-blower. "Consequently, we have a dozen Fukushimas waiting to happen in America."

THE MELTDOWN IN JAPAN COULDN'T have happened at a worse time for the industry. In recent years, nuclear power has been hyped as the only energy source that could replace coal quickly enough to slow the pace of global warming. Some 60 new nukes are currently in the works worldwide, prompting the industry to boast of a "nuclear renaissance." In his 2012 budget, President Obama included $54 billion in federal loan guarantees for new reactors—far more than the $18 billion available for renewable energy.

Without such taxpayer support, no new reactors would ever be built. Since the Manhattan Project was created to develop the atomic bomb back in the 1940s, the dream of a nuclear future has been fueled almost entirely by Big Government. America's current fleet of reactors exists only because Congress passed the Price-Anderson Act in 1957, limiting the liability of nuclear plant operators in case of disaster. And even with taxpayers assuming most of the risk, Wall Street still won't finance nuclear reactors without direct federal assistance, in part because construction costs are so high (up to $20 billion per plant) and in part because nukes are the only energy investment that can be rendered worthless in a matter of hours. "In a free market, where real risks and costs are accounted for, nuclear power doesn't exist," says Amory Lovins, a leading energy expert at the Rocky Mountain Institute. Nuclear plants "are a creation of government policy and intervention."

They are also a creation of lobbying and campaign contributions. Over the past decade, the nuclear industry has contributed more than $4.6 million to members of Congress—and last year alone, it spent $1.7 million on federal lobbying. Given the generous flow of nuclear money, the NRC is essentially rigged to operate in the industry's favor. The agency has plenty of skilled engineers and scientists at the staff level, but the five commissioners who oversee it often have close ties to the industry they are supposed to regulate. "They are vetted by the industry," says Robert Alvarez, a former senior policy adviser at the Energy Department. "It's the typical revolving-door story—many are coming in or out of jobs with the nuclear power industry. You don't get a lot of skeptics appointed to this job."

Jeffrey Merrifield, a former NRC commissioner who left the agency in 2007, is a case in point. When Merrifield was ready to exit public service, he simply called up the CEO of Exelon, the country's largest nuclear operator, and asked him for a job recommendation. Given his friends in high places, he wound up taking a top job at the Shaw Group, a construction firm that builds nuclear reactors—and he's done his best to return the favor. During the Fukushima disaster, Merrifield appeared on Fox News, as well as in videos for the Nuclear Energy Institute, the industry's lobbying group. In one video—titled "Former NRC Commissioner Confident That Building of New U.S. Nuclear Plants Should Continue"—Merrifield reassures viewers that the meltdown in Japan is no big deal. "We should continue to move forward with building those new plants," he says, "because it's the right thing for our nation and it's the right thing for our future."

Such cozy relationships between regulators and the industry are nothing new. The NRC and the utilities it oversees have engaged in an unholy alliance since 1974, when the agency rose from the ashes of the old Atomic Energy Commission, whose mandate was to promote nuclear power. "For political reasons, the U.S. wanted to show something good could come out of splitting the atom," says Robert Duffy, a political scientist at Colorado State University who has written widely about the history of nuclear power. "There was great pressure on the industry to get nuclear plants built quickly." With no effective oversight by the government, the industry repeatedly cut corners on the design and construction of reactors. At the Diablo Canyon plant in California, engineers actually installed vital cooling pipes backward, only to have to tear them out and reinstall them.

But even the lax oversight provided by the NRC was more than the industry could bear. In 1996, in one of the most aggressive enforcement moves in the agency's history, the NRC launched an investigation into design flaws at a host of reactors and handed out significant fines. When the industry complained to Sen. Pete Domenici of New Mexico, a powerful nuclear ally, he confronted the head of the NRC in his office and threatened to cut its funding by a third unless the agency backed off. "So the NRC folded their tent and went away," says Lochbaum. "And they've been away pretty much ever since."

THE JAPANESE DISASTER SHOULD HAVE been a wake-up call for boosters of nuclear power. America has 31 aging reactors just like Fukushima, and it wouldn't take an earthquake or tsunami to push many of them to the brink of meltdown. A natural disaster may have triggered the crisis in Japan, but the real problem was that the plant lost power and was unable to keep its cooling systems running—a condition known as "station blackout." At U.S. reactors, power failures have been caused by culprits as mundane as squirrels playing on power lines. In the event of a blackout, operators have only a few hours to restore power before a meltdown begins. All nukes are equipped with backup diesel generators, as well as batteries. But at

Fukushima, the diesel generators were swamped by flood-waters, and the batteries lasted a mere eight hours—not nearly long enough to get power restored and avert catastrophe. NRC standards do virtually nothing to prevent such a crisis here at home. Only 11 of America's nuclear reactors have batteries designed to supply power for up to eight hours, while the other 93 have batteries that last half that long.

And that's just the beginning of the danger. Aging reactors are a gold mine for the power companies that own them. Nuclear plants are expensive to build but cheap to operate, meaning the longer they run, the more profitable they become. The NRC has done its part to boost profitability by allowing companies to "uprate" old nukes—modifying them to run harder—without requiring additional safety improvements. Vermont Yankee, for example, was permitted to boost its output by 20 percent, eroding the reactor's ability to cool itself in the event of an emergency. The NRC's own advisory committee on reactor safety was vehemently opposed to allowing such modifications, but the agency ultimately allowed the industry to trade safety for profit. "The NRC put millions of Americans at elevated risk," says Lochbaum.

Indeed, the NRC's "safety-last" attitude recalls the industry-friendly approach to regulation that resulted in the BP disaster in the Gulf of Mexico last year. Nuclear reactors were built to last only 40 years, but the NRC has repeatedly greenlighted industry requests to keep the aging nukes running for another two decades: Of the 63 applications the NRC has received for license extensions, it has approved all 63. In some cases, according to the agency's own Office of the Inspector General, NRC inspectors failed to verify the authenticity of safety information submitted by the industry, opting to simply cut and paste sections of the applications into their own safety reviews. That's particularly frightening given that some of America's most troubled reactors—including Davis-Besse in Ohio, where a football-size hole overlooked by NRC inspectors nearly caused a catastrophe in 2002—are now pushing for extensions. "If history is any judge, the NRC is likely to grant them," says Gundersen, the former nuclear executive.

Even after a reactor is found to be at higher risk because of new information about earthquake zones—as is the case at Indian Point, located only 38 miles from New York City—the NRC has done little to bolster safety requirements. The agency's current risk estimate of potential core damage at the Pilgrim reactor in Plymouth, Massachusetts, is eight times higher than its earlier 1989 estimate—yet it has done little to require the plant to prepare for an earthquake, beyond adding a few more fire hoses and other emergency gear. The Diablo Canyon plant in California, which sits near one of the most active seismic zones in the world, is supposedly engineered to withstand a 7.5 earthquake. There's only one problem: Two nearby faults are capable of producing quakes of 7.7 or higher. Should it be shut down? "That's the kind of big question the NRC should be capable of answering," says Gilinsky, the former NRC commissioner. "Unfortunately, they are not."

The biggest safety issue the NRC faces with old nukes is what to do about the nuclear waste. At Fukushima, the largest release of radioactivity apparently came from the concrete pools where spent fuel rods, clad with a special alloy, are placed to cool down after they are used in the reactor. These spent rods are extremely hot—up to 2,000 degrees Fahrenheit—and need a constant circulation of water to keep them from burning up. But in America, most plants have no way of keeping the water circulating in the event of a power failure. Nor are the pools themselves typically housed in secure bunkers, because the NRC long considered it virtually impossible for the special alloy to catch fire.

Fukushima proved them wrong. The earthquake damaged the systems that cooled the spent rods, allowing the water to drain out. The rods then heated up and the cladding caught fire, releasing cesium-137 and other radioactive particles. The rods were eventually cooled with seawater fired from water cannons and pumped in by firetrucks, but not before a significant amount of radiation had been released.

In theory, pools in the U.S. were only supposed to hold spent fuel rods for a short time, until they could be moved to a permanent disposal site at Yucca Mountain in Nevada. But the site has remained unfeasible despite two decades and $7 billion in research, prompting President Obama to finally pull the plug on it last year. That means tens of thousands of tons of irradiated fuel continue to sit in spent fuel pools at reactors across the country—America's largest repository of radioactive material. A release of just one-tenth of the radioactive material at the Vermont Yankee reactor could kill thousands and render much of New England uninhabitable for centuries. "Yet the NRC has ignored the risk for decades," says Alvarez, the former Energy Department adviser.

According to a 2003 study, it would cost as much as $7 billion to move the spent fuel out of the pools and into more secure containers known as dry-cask storage. So why hasn't the NRC required such a precaution? "Power companies don't want to pay for it," says Alvarez. "They would rather let the public take the risk." Gilinsky offers another explanation. "After insisting for years that spent fuel pools were not a problem," he says, "the NRC doesn't want to admit what everyone knows after Fukushima: They were wrong."

AS CHAIRMAN OF THE NRC, Gregory Jaczko was supposed to reform the agency. He formerly served as science adviser to Sen. Harry Reid of Nevada, and won his seat on the commission in 2005 over protests from the industry. Under his leadership, however, the NRC has displayed an alarming lack of urgency in the wake of Fukushima. The agency says it is currently taking a quick look for immediate problems at U.S. reactors, and promises to follow up with a more in-depth review later. But it's an indication of how little respect the agency commands that no one expects much to change. Indeed, ever since the terrorist attacks in 2001, the NRC has become increasingly secretive. "The agency has used national security as an excuse to withhold information," says Diane Curran, an attorney who specializes in nuclear safety.

Some critics argue that it's time for an outside agency, such as the National Academy of Sciences, to take an independent look at the safety and security of America's aging nukes. A better idea might be to simply repeal the Price-Anderson Act and force the nuclear industry to take responsibility for the risks of running these old plants, rather than laying it all off on taxpayers. The meltdown in Japan could cost Tokyo Electric some $130 billion—roughly three times what the Deepwater Horizon spill cost BP. If nuke owners had to put their own money where their atoms are, the crumbling old reactors would get cleaned up or shut down in a heartbeat.

Instead, by allowing the industry to cut safety margins in exchange for profits, the NRC is actually putting the "nuclear renaissance" itself at risk. "It has not been protesters who have brought down the nuclear industry," said Rep. Ed Markey of Massachusetts. "It has been Wall Street." Wind and natural gas are already cheaper than nukes, and the price of solar is falling fast. And with each new Fukushima, the cost of nukes—as well as the risks—will continue to rise.

"The question is not whether we will get an earthquake or a tsunami," says Lochbaum. "The question is whether we are fully prepared for unexpected events, and whether we are doing everything we can to protect the public. I don't think we are. If and when there is a nuclear disaster, I would hate to be the one who has to stand up in front of the American people and say, 'We knew about these problems, but did nothing about them.'"

P. J. O'ROURKE

The Last Shuttle Launch

FROM *THE WEEKLY STANDARD*

With the shuttle program over, and NASA focusing on other priorities, P. J. O'Rourke laments the end of a heroic era of manned space exploration.

MERRITT ISLAND, F.L.—MY SEVEN-YEAR-OLD SON, Cliff, watched the last space shuttle launch from the NASA viewing stands at the Kennedy Space Center. He had a spiritual experience of a kind that no amount of dragging him to Mass or even Fenway Park has inspired. His little face—seemingly made up entirely of open eyes—announced it: *"This* is awe!" He didn't need to say anything and, having forgotten to breathe, he probably couldn't. Indeed, for the first waking moment in his 89 months on earth, he was silent.

The swooped delta of the *Atlantis* shuttle with its orange gothic squid of a liquid fuel tank and its twin column, party hat-topped solid-fuel boosters—the "full stack" as it's called—is three-and-a-half miles away but looms nonetheless. Perching on its launcher it is as tall as a 25-story building. There's a flash below the engine nozzles. A fiery glory pours out on every side. A few seconds later comes the joyful noise, a trumpeting so powerful that the decibels will kill you if you're closer than 800 feet.

My eyes were as wide as my son's, but, unlike him, I was babbling as if I were a blind, deaf, and dumb man miraculously cured: "*This* is light!" "*This* is sound!"

The full stack stands almost still, trembling with the strength of 6,825,704 foot-pounds of thrust. Then it is risen—ascending on a tower of smoke with the slow majesty befitting 2,030 tons of wondrous engineering. The *Atlantis*, still joined to its external tank and boosters, rolls gracefully onto its back, embracing the heavens now instead of the Earth, and traces an arc as grand as the curve of the space-time continuum. Then it disappears into the layer of stratocumulus that had been threatening for three days to scrub the launch.

Cliff made a small noise of protest. With a child's love of things coming apart, he'd wanted to see the solid boosters detach and fall into the ocean. But there was a better show. Rocket smoke cast a scything shadow out to the horizon across the cloud tops while at the launch site smoke still stood, vertical, immense, and undissipating. "He took not away the pillar of cloud by day."

And He didn't. Congress and the White House did. Since the end of the countdown about two minutes had passed, which seemed like an era, and it was—the end of an era. That's that for the NASA shuttle program and maybe for the whole idea of U.S. manned space exploration.

We used to have presidents who liked to send Americans places—Iraq, Afghanistan, the Moon, or Mars. But George W. Bush's NASA Constellation program has been canceled. Its gigantic Ares V rocket is off the drawing board. The Constellation's *Orion* flight capsule has been renamed, in a telling translation into GovSpeak, MPCV—"Multi-Purpose Crew Vehicle." What the multiple purposes are supposed to be is anyone's guess. At the moment the only way NASA can get a person into space is by paying Vladmir Putin for a ride on the creaky old *Soyuz*. Looks like the Russians won the space race after all. Meanwhile America's government has not funded, or even proposed, anywhere new for people to go beyond low Earth orbit. Never mind that the observable universe is 92 billion light years across and would seem to offer ample travel opportunities.

In a rare outburst of bureaucratic blunt truth, Michael Leinbach, the *Atlantis* launch director, told colleagues at the Kennedy Space Center, "We're all victims of poor policy out of Washington. I'm embarrassed that we don't have better guidance."

On the trip to Florida I explained America's space exploration to Cliff as well as I could. In a sense he understands better than I do. To him space travel is not an extraordinary phenomenon but a long-standing historical movement, an inevitability in the course of human affairs. It's what the discovery and settlement of the American West was to me when I was his age, before Manifest Destiny was insulted in grade schools.

Cliff's reaction to the news that America will now lack a manned spaceship was like hearing from Miss Sonnenberg, my second grade teacher, that America had stopped with the Louisiana Purchase; there was no war with Mexico, and the heck with Texas and California.

Cliff and I went to the shuttle launch through the good offices of our friend G. Ryan Faith, research analyst at the Space Foundation, a nonprofit that brings together everyone involved in space exploration—civilians, the military, commercial entrepreneurs, and government space agencies from around the world. I think aliens from Area 51 in Nevada would be welcome if they existed. One purpose of the foundation is to build understanding, support, and enthusiasm for what would be the coolest thing in the world except all sorts of other worlds are involved so it's even cooler than that.

Places for children in the NASA viewing stands are hard to come by, reserved mostly for astronaut offspring. But I argued that if you're going to promote the coolest thing in the world you need testimony from someone who really uses those words in daily speech. The Space Foundation put us in touch with the John H. Glenn Research Center, which is, among other things, NASA's main propulsion test facility, and let's hope they have something to propel soon. James M. Free, the deputy director, extended invitations, and Cliff became the National Aeronautics and Space Administration's most vocal advocate—after his power of speech returned.

Two nights before the launch the Glenn Center hosted a reception in Cocoa Beach. Actual astronauts were in attendance, including Rick Mastracchio, who has been on three shuttle missions and done at least three space walks. He was wearing his blue NASA flight suit. I introduced Cliff, and Cliff came right to the point about the cancellation of the shuttle program. Well, not right to the point. He had some questions of greater seven-year-old importance to ask first. "When you were weightless, did you do a flip?"

"I sure did," Mastracchio said.

"Did you do a 'Misty'?" asked Cliff.

Mastracchio looked puzzled. "A snowboarding term," I explained. "A back flip with a 540-degree spin."

"Good idea," said Mastracchio. "I'll try it next time."

But Cliff knew there won't be a lot of next times. Only four Americans a year will fly on the *Soyuz*. A decade ago the astronaut corps had about 150 men and women; now there are 61.

Cliff said, "Why are they putting the shuttle down?"

It was a nice choice of phrase. A country boy knows what happens to old or unwanted animals, and he combined that with the term for a schoolyard diss—the space geeks being razzed by the popular kids from the Oval Office and the Hill.

Mastracchio answered diplomatically. NASA is working on a new heavy-lift rocket. (Though it probably won't be ready until 2020, leaving budget-cutters plenty of time to work.) Private commercial manned flights are on the way. (But not before 2016, and only if Congress is willing to pay for seats on board.) "By the time you're ready to be an astronaut," Mastracchio said, "there will be plenty of ways to go to space."

Cliff was not convinced. "Is the government," he asked, "just being mean?"

A significant glance was exchanged between astronaut and parent over Cliff's head.

So enough's enough, Christopher Columbus. Four voyages were plenty. The natives are crabby. There's no gold out there. And Don Quixote needs meds. Sancho Panza's pension plan has to be fully funded. A wide variety of social and educational initiatives are necessary for Dulcinea to achieve her full potential as a wench. Anyway, if we need to go to the New World again, Sir Francis Drake would consider having a Spaniard onboard to be a real prize.

On July 20, 1969, at 10:56 p.m. I was in my off-campus apartment staring at a black and white portable TV, a can of Budweiser in my hand. "One small step . . . " I remember every detail. Where were you and what were you doing when Lyndon Johnson signed Medicare into law? Medicare cost $523 billion in 2010. NASA cost $18.7 billion, just 0.6 percent of federal spending. In fact, since NASA's founding in 1958, its total spending has barely exceeded what we pay for Medicare per annum. Would you rather reach into infinity for 53 years or get old and sick for 12 months? In 2011 each American will give NASA about $60—the sun, the moon, and the stars for less than the price of a month of basic cable.

Oh, maybe it's a waste of taxpayer money. But government wastes taxpayer money. This is what government does. It can't be changed. Our earliest evidence of government, in the ruins of Babylon and Egypt, shows nothing but ziggurats and pyramids of wasted taxpayer money, the TARP funds and shovel-ready stimulus programs of their day. Let's waste taxpayer money putting that look back on Cliff's face.

ERIK SOFGE
_____

The Early Adopter's Guide to Space Travel

FROM *POPULAR MECHANICS*

> With commercial space travel on the horizon, voyagers may want to know what might be in store when they book a flight beyond Earth's orbit. Erik Sofge provides a preview.

2012—THERE ARE DOZENS OF MILES TO GO BEFORE YOU'RE an astronaut. But at an altitude of 50,000 feet, your spaceflight is under way. That's when WhiteKnightTwo, the plane that has been in control so far, releases your craft like a massive bomb, and it starts to free-fall. After a 3000-foot drop, SpaceShipTwo's rocket fires. The craft bucks. The thrust presses you deep into your seat, building as you climb. On paper, you're experiencing nearly 4 g's, equal to four times Earth's gravity. In practice, this gradual ramp-up to Mach 3 is smooth and exhilarating. The media might focus on the weightlessness you'll soon feel, but that's not the point. You don't spend up to six years on a Virgin Galactic waiting list, or longer than that, maybe an entire lifetime, dreaming of floating. The dream was always to land a seat on a rocket ship.

Not everyone shares this dream, or even respects it. Decades ago, the world watched in wonder as the first NASA astronauts escaped the atmosphere to circle the planet and later set boots on the moon. Today, being one of the first 100 customers to put down $200,000 for a trip into space has earned you a place in Virgin Galactic's founders group. From the wider world, that honor has inspired a collective shrug. After all, starting with Dennis Tito's $20 million ride aboard a Russian Soyuz in 2001, a handful of citizen astronauts flew earlier and higher. Tito and the six Space Adventures customers who followed him made it all the way to orbit, where they boarded the International Space Station. The craft that Virgin Galactic flies, SpaceShipTwo, barely kisses the void at a peak altitude of 68 miles, offering just 4 to 5 short minutes of weightlessness before gravity pulls it back into the atmosphere. It may be a giant leap for the leisure class, but for mankind's expansion into space, it's hard to imagine a smaller baby step.

Of course, no one signs up for the amenities: There are no flight attendants passing out champagne or directing your attention to the bathrooms. Rocket ships don't have flight attendants—or bathrooms. They have rockets, which are sinking those 4 g's straight into your chest as the blue sky outside your window drains to black. Part of you hopes the ascent never ends—this amount of power seduces. In fact, many of Virgin Galactic's customers are pilots or "people who simply love flight," says CEO George Whitesides.

SpaceShipTwo is lean and efficient, more like an experimental aircraft than a commercial plane. A special mother ship hauls it into the sky, then hybrid rocket motors take over; the craft returns to the runway in a controlled glide. It is the direct descendant of SpaceShip-One, which took the $10 million Ansari X Prize by making two suborbital flights within two weeks in 2004. Virgin Galactic partnered with the vehicle's developer, Scaled Composites, to create a follow-up to the winning design, and after years of development, SpaceShipTwo was unveiled in 2009. It's carried aloft by White-KnightTwo, and along with seats for a pilot and co-pilot, it has enough room for six passengers to briefly unbuckle and drift around the cabin.

Virgin Galactic isn't the only company with suborbital aspirations in 2012. XCOR Aerospace plans to undercut Virgin with $95,000 flights on its two-person Lynx space plane; tickets for Armadillo Aerospace's vertical launching and landing rocket will start at $100,000. According to Whitesides, Virgin Galactic is already looking past suborbital, throwing its support behind the development of orbit-class shuttles from both Sierra Nevada Space Systems and Orbital Sciences Corp.

ALTHOUGH THE PRIVATE SPACE INDUSTRY is growing, it's far from viable, or proven. That's where you come in. Someday they'll say that mass space travel started in 2012. Before it extended into orbit and beyond, the suborbital hops sustained a global industry. And you were there, as a deep-pocketed early adopter, a zero-gravity guinea pig, a philanthropist donating to a dream of private spaceflight that sounds like science fiction and might never be profitable. Most of all, though, you were there as an astronaut—or a spaceflight participant, according to the Federal Aviation Administration (FAA)—on SpaceShipTwo.

The transition to nearly zero g, or microgravity, happens quickly, and at an altitude of 62 miles, your float time is under way. Here's where your training matters. You've been in microgravity before, on multiple parabolic flights, where 727s perform an undulating series of swoops and soars, creating something like weightlessness for up to 30 seconds at a time. The sensation is sickening for a small percentage of people, but for most, it tends to focus all attention on the suddenly novel mechanics of motion in a gravity-free environment. When passengers aren't giggling, they're moving all wrong, trying to swim in midair or pushing too hard off walls and objects. It takes a little practice to turn the volume down, stop being amazed at this altered state and remember to look out the window.

The window is where you're headed. Not to take a picture—the craft is studded with cameras inside and out, to discourage participants from frittering away precious time squinting into their own point-and-shoots—but to see what's out there. You push off the seatback, keeping your legs and arms tucked close. Everyone on board has some sort of plan, and they all start the same, with a drift toward one of the wide, circular windows scattered liberally throughout the craft. The 5-minute countdown is ticking away, and it's nonnegotiable. When SpaceShipTwo's arc turns into a fall, it falls hard, peaking at 6 g's. That's potential blackout territory, and serves as the craft's effective flight ceiling—any higher and passengers would stay weightless longer but face even more g-loading upon re-entry. You prepared for that too: During your three-day preflight training at New Mexico's Spaceport America, you practiced the breath-clenching tricks fighter pilots use to keep from blacking out during high-speed maneuvers. That's the part of the trip, the ride down, that will test your mettle.

The descent will come in a few minutes. Right now, you are 68 miles high. The Earth fills your view. It's all that matters.

2022—IT IS CALLED THE OVERVIEW effect. It's not a single epiphany, but rather a series of them, all triggered by the simple act of staring at the world from space. After a full week in orbit, it's been a while since your last big one. From an earthward window in the habitat, you watch a fire blooming somewhere in the Midwest, along the Michigan-Ohio border. That was an early realization—that something is always burning. And that you can always see it, even when the smoke from a single burning house rises in a column and spreads, pooling against the dome of the atmosphere to cover a whole state. Earth-gazing is one of the few luxuries aboard a space station. Of the handful of cylindrical habitats docked together to form this orbital space station—each one launched individually by Bigelow Aerospace, inflated in orbit and steered into place—none holds a floating spa. Even as more habitats are added, the station expanding like a massive balloon animal, most will hold sleeping quarters and laboratory space.

Maybe one day a hospitality firm will erect a true space hotel— little clusters of private rooms with body-length windows and gourmet cuisine that doesn't ooze out of pouches. But for you, it's back to the grind. After all, no one goes to space to look at Earth all day. Most people here are flown into space on their companies' dime—in your case, to create protein crystals that grow purer and larger in microgravity for more precise drug testing. For the foreseeable future, space will be a workplace.

You fought for your spot, but for most of the thousands willing to risk that climb aboard a three-stage, orbit-class spaceship, the costs are prohibitive. This has almost nothing to do with fuel—which can account for as little as 0.5 percent of an orbital flight's price tag—and a whole lot to do with disposable vehicles. Even the most cost-effective rockets, such as SpaceX's Falcon 9, are ditched midflight. "A 747 costs around a quarter-billion dollars," says SpaceX CEO Elon Musk. "It doesn't cost half a billion dollars to fly round trip. That's because you don't destroy them after one use." Musk estimates that true reusability—a double-stage rocket design, or boosters that could fly themselves home—would lower costs by a factor of more than 100. However, it would also be "one of the most difficult and important inventions that humanity will ever achieve," he says.

In 2022, rockets are cheaper but far from reusable. So, while suborbital travel has exploded over the past decade, with a few thousand flights per year and prices settling in the $20,000 to $30,000 range, the orbital market is dominated by research-minded corporate and government clients that are willing or able to pay many millions of dollars for a week or two in space. Even the visitors who pay their own way aren't really tourists; they float through these same lab modules with their own for-profit research projects or experiments. "We know that anyone we're going to bring up, even for a short flight, won't just want to float around and take pictures," says former astronaut Leroy Chiao, today an executive vice president at Excalibur Almaz, which has purchased Russian space station modules for potential deployment. "Pretty soon that gets old. The people doing this want to do something meaningful," Chiao says.

Your time in space is running out. Tomorrow—in 24 hours, that is, since here the sun rises every 90 minutes—you'll be leaving, heading home in a shuttle that's scheduled to arrive at the station any minute. It's the same kind you rode up in: a Dream Chaser, carrying six passengers, one pilot and all the research gear and resupply cargo that can fit in the hold. The trip home, you've heard, is smoother than many suborbital flights, since the craft angles gradually into a 1.6 g's re-entry rather than diving in a relatively steep decline. The hardest part of orbital travel, it's said, is orbiting, and all the strange forces this environment exerts on the human body.

The most obvious effects include fluids shifting, which raises blood pressure as well as pressure in the lungs, face and eyes. For some, sleep can be difficult. And for most, there is some level of motion sickness. Richard Garriott, a computer-game designer and the sixth civilian astronaut, says roughly 80 percent of people feel sick in orbit, and pilots and roller coaster enthusiasts don't get a pass. "There's no correlation we can find between motion sickness on Earth and motion sickness in microgravity," says Garriott, who is Space Adventures' largest shareholder. It was his trip to the ISS in

2008 that cleared astronauts who'd had laser eye surgery to go or-bital. Until then, doctors had feared elevated pressure might tempo-rarily distort their vision.

It was a physical ailment that helped secure your spot on the space station. There were plenty of other applicants at your pharmaceuti-cal company, most of whom could conduct the same research. But you happen to have Type 1 diabetes. The opportunity to collect data on microgravity insulin injections and the day-to-day fluctuations of your blood sugar tipped the scales. These are the early days of the orbital-data gold rush, when a single procedure or study is often valued in the millions. Each flight is a major investment, and com-panies like yours want to squeeze the most out of every trip.

More of those investments are inbound. An announcement over the PA system sends you back to the window. The sun is rising, again, and the Dream Chaser's white surface is almost too bright to look at. Its Falcon 9 booster is long gone, and the shuttle—a smaller-winged, sawed-off cousin of NASA's—is slowly rotating into a dock-ing position. Behind it, you see more fires on Earth's surface. This isn't the Midwest, though. It's the Amazon rain forest, which has been burning here and there since you first reached orbit. For de-cades, in fact, astronauts have watched that land smolder.

The orbital view is harsh and matter-of-fact. There is beauty below, but humans are everywhere, their cities and suburbs filling the most livable spaces, the indelible signs of industry stitched across the others. This is your final epiphany: It's not a big world at all. It's smaller than you can possibly describe.

2042—IT'S NOT A TRAFFIC JAM, but there is congestion in orbit. The pilot nudges your craft into a slow, slightly woozy spin as you stand by for clearance from the FAA. You don't mind the wait. You've never been on an old-fashioned rocket ship, with a massive liquid-fuel booster trailing the crew vehicle. There's a lot of potential energy sitting behind you—enough, it seems, to make it to the pocked, luminous rock now wheeling into view.

Ideally, there would be a countdown, an old NASA-style voice crackling T-minus updates over the radio. But compared to what Apollo astronauts experienced, your blastoff to the moon from 200 or so miles in orbit is a relatively casual affair. The pilot seated next to you in this well-appointed two-person shuttle confirms with traffic controllers that your flight path won't intersect with the orbit of the half-dozen habitats hurtling around the planet, and you almost hope he misses the launch window. If he did, you would spend an extra hour circling Earth, sitting in the cockpit of an almost comically small spaceship perched atop a mammoth booster vehicle at least 10 times its size. That would make this feel more like a mission and less like a joyride.

You wonder whether anyone is watching from one of the habitats or from Earth itself as the booster attached to the back of this shuttle fires. Probably not. Lunar fly-bys are almost universally ignored and generally considered a costly, pointless side trip for the idle rich, who have all but run out of adventures in space. Besides, privately funded rovers have been ambling through lunar craters for years, and a handful of civilians have already looped around the moon, back when that kind of a mission was, you know, hard.

Of course, nothing in space is actually easy. The shuttle you're in, which is similar to the Lynx suborbital space plane that used to fly four times daily or as often as 800 hops per year, is nothing like the cramped, no-frills Soyuz capsule that Space Adventures managed to swing around the moon many decades after Apollo. Instead of peeking through portholes, you have a panoramic view of the stars and the approaching lunar surface. An additional, miniature habitat module sits behind your shuttle so that you can spread out during the five-day trip. What makes this flight so expensive isn't the duration or the training required, which are almost identical to those of an orbital stay. It's the sheer energy needed to travel 240,000 miles and back, supplied by the booster mated to your craft.

This kind of ticket once cost as much as $150 million, with spaceflight participants footing the bill for a colossal amount of gear to be hauled into space: a crew vehicle and either a pair of rockets to take that craft orbital and then around the moon, or a single, larger rocket that could do both. In 2042, the price is in the low millions, thanks to a series of breakthroughs. The biggest of them was the strange rise of the beamship. This radical redesign of orbit-class spacecraft started small, with tiny, unmanned vehicles powered by ground-based microwave arrays. The beams would track a craft's hot-plate-like underside as it flew, heating and expanding the hydrogen onboard, producing more high-velocity thrust than traditional oxygen-and-hydrogen-burning rockets. The advantages of beampowered flight are cumulative: Smaller, simpler, more stable craft would also be inherently lighter, and the heat exchanger that serves as a microwave target could act as a heat shield during re-entry.

"I know this sounds like science fiction the first time you hear about it, but when you look at the actual details, the numbers, it's not all that far-fetched," says Kevin Parkin, who heads Carnegie Mellon's Microwave Thermal Rocket project in collaboration with NASA.

Scaling up from robotic resupply missions to cheap, booster-free manned flights was inevitable, and brought the price of an orbital ticket down below $100,000. In 2042, beamships regularly ferry cargo and passengers to orbit, while the purely chemical rockets and capsules are being retooled for manned missions to asteroids, Mars and other deep-space destinations. Every so often, a booster is tested by pushing a beamship around the moon—in this case, yours.

A couple of days into the flight, your lunar loop is nearing the halfway mark. Back in the cockpit, the pilot points out major features as you fly 62 miles above the moon's surface, but mostly this leg of the trip is quiet. You're silent, watching craters flow by and imagining how long it will take, how many more flights like this, before boosters will be completely reusable, able to regularly land on that craggy, alien landscape, tank up with fuel harvested from automated lunar mines and lift off again. How long until the moon is teeming with life instead of dotted with roving machines. When, you wonder, will space really open up?

Then Earth rises from behind the moon, bright and bluer than any picture, any video or any description can capture, a glow that seems to spill into the cracked, lifeless lunar bowls and valleys below. It's as beautiful as those Apollo astronauts claimed, so pretty it's almost embarrassing.

Even so, your mind is already skipping ahead. Someone must be booking a flight around Mars. You should look into that.

CHARLES PETIT

Stellar Oddballs

FROM *SCIENCE NEWS*

> *NASA launched the Kepler space telescope to find Earth-like planets orbiting faraway stars. As Charles Petit reports, Kepler is turning up so much unexpected data on stars and other exotic objects that astrophysicists are scrambling to keep up.*

AFTER MIND-BENDINGLY PRECISE DATA AND ARTISTS' renditions of mysterious stars played across the screen, Martin Still leaned into his lectern at an American Association for the Advancement of Science meeting early this year to deliver a plea to fellow astronomers. In one word: Help!

"We need you guys," said the manager of NASA's guest observer program for Kepler, among the most successful space telescopes ever launched. "Wait a year and it's too late."

Kepler has found a bonus, a treasury of wonders, or one might say a stellar freak show out in space. The result is a predicament: This is not what the space telescope was looking for. Kepler's NASA team has one job that it must, by contract, pursue almost exclusively—hunting for extra-solar planets. In-house researchers must largely ignore other wonders. Hence the call for aid from guest observers, people given access to a telescope's data, but who typically provide their own resources to analyze the results and pursue more.

In science, new instruments routinely discover unexpected things. But Kepler's surprises—which could help astronomers learn far more about the evolution of stars, their internal structures and how the burning balls of plasma die—require fast action if they are to be fully examined.

The telescope honors 17th century German astronomer Johannes Kepler. He was first to realize that planets follow elliptical, not circular, orbits, and he established three laws of planetary motion. "Our mission is to find planets. We hope to find Earthlike planets," says the project's founder and principal investigator, William Borucki of NASA's Ames Research Center in Mountain View, Calif. Borucki spent decades fighting against great skepticism to build an orbiting instrument so sensitive it would detect planets that briefly cross, or transit, their stars' faces.

NASA launched the telescope in March 2009 into a "trailing" orbit. Sitting slightly farther from the sun than Earth does, Kepler makes one trip around the sun every 372 days, gradually falling farther behind Earth. In its solitude, the spacecraft keeps its eye on hordes of stars, checking for planets.

It is finding them, too, in scads. There are more than 1,200 entries on Kepler's list of candidate planets betrayed by small, repetitive and distinctively shaped dips in the brightness of their parent stars—the shadow of a planet crossing a star's face as seen by the craft. While further analysis surely will reveal some as false alarms, chances are that 90 percent are real. That will triple the number of known exoplanets. With time, scientists hope to confirm a few with Earthlike size, orbit and other conditions suitable for life as we know it to arise.

But the majority of stars don't have planets lined up to block light headed toward Kepler. These stars too are worthy of study. Still, who is based at Ames, hopes outside astronomers will take a close look at Kepler's new data on stars without signs of planets, plus look at those stars with other scientific instruments while the craft is still operational.

EYE FOR THE WEIRD

"There are so many stars that show bizarre, utterly unexplainable brightness variations that I don't know where to begin," says Geoff Marcy of the University of California, Berkeley. Marcy gained fame in the mid-1990s when he helped pioneer, with ground-based instruments, discovery of extrasolar planets; he joined the Kepler team to help expand the planet-finding toolbox. After the detection of far more than planets began overwhelming the Kepler program, Marcy hired an undergraduate statistics major to scan the plots of varying brightness for tens of thousands of stars. He trained her to spot what Marcy calls the WTF objects, which might politely be rendered "What The Flip is that?"

Astronomers have already spotted stars with remarkable pulsation modes, double stars orbiting so closely that streams of white-hot plasma flow between them, immense star spots whose movements hint at unlikely rotations, collapsed white dwarf stars in eclipsing orbits around large and seemingly younger stars, and more. "These phenomena have never been seen before, or never with such clarity. This is a gold mine," Marcy says.

Ironically, the bounty comes from a telescope that Borucki has often called, tongue in cheek, the most boring space mission in history. All Kepler does is look, occasionally pausing to realign its solar panels, at one starry patch of sky about as big as a hand held at arm's length. The target is roughly between the Northern Hemisphere constellations Cygnus and Lyra.

About 4 million stars in Kepler's view are bright enough to study closely. Unlike most telescopes, its camera gathers no spectra that directly reveal chemical composition. The main objective is to measure star brightness, an astronomical procedure called photometry. Stellar photometry has never been done before on this scale or with this accuracy. With 42 CCD photo chips, or about 95 megapixels in all, Kepler's camera—the largest ever put in orbit—can detect changes in brightness to well under 0.01 percent, good enough for Earth-sized planets orbiting sun-sized stars. There is capacity on the radio link to Earth for about 6 percent of the data avalanche. "We throw away the rest," Still says. Thus choosing which stars' data to hold onto is essential.

Most of Kepler's attention is on 156,000 stars selected because they appeared from pre-Kepler data to be more or less sunlike, fairly stable in their output. Even among these stars, most light curves are not checked by human eye; computer programs recognize and flag those with the regular, slight dips suggestive of planets. Researchers can request up to 512 extra objects for detailed study: stars already excluded from planet study, perhaps, but interesting for other reasons, or background galaxies.

Already, general astrophysics papers (meaning papers about stars) published from the mission outnumber those dealing with planets. Some papers are on single stars that turned out to be double or triple systems, or stars orbited by—and sometimes consuming—whirling disks of gas, plasma and dust. Such things were already known to exist, but to have so many in such detail is new. Anybody can go to the Kepler website, pick at random one of the 156,000-plus or so stars on the target list, and command the Kepler server to plot and display the star's light curve.

An Amateur Helps Himself

Even amateurs are welcome to try. One such amateur is Kevin Apps, an engineer in Surrey, England. He works days managing natural gas pipeline flows but has published several astrophysics papers with professionals. Shortly after Kepler went into orbit, Apps looked in the telescope's data file for a red dwarf about 120 light-years away that he knew about from star catalogs. To his surprise it had not been included in the list of 156,000 for Kepler's close attention.

From his home computer Apps found that he could retrieve the light curve using data gathered during the telescope's initial commissioning phase. The light curve had four dips spaced 12.71 days apart, suggestive of planet-sized transits. He contacted John A. Johnson, an assistant professor at Caltech whom Apps had worked with before.

Intrigued, Johnson recruited eight other professional astronomers. They obtained spectra with telescopes in California and Hawaii. Analysis revealed not, as the catalogs say, a single star but instead a wide-spaced pair, or binary, of red dwarfs (also known as M dwarfs, the most common kind of star in the galaxy). One is about 30 percent and the other 37 percent as massive as the sun. They circle one another every 100 years or so. With an adaptive optics system on the famed 200-inch telescope at Mount Palomar in California, the team even got a fuzzy photo showing two distinct stars.

And the dips in starlight? They are indeed caused by something about the size of a large planet in orbit about the larger of the two dim stars. But it is no planet. Orbital motion betrays a mass 63 times that of Jupiter, Johnson, Apps and colleagues reported April 1 in the *Astrophysical Journal*. That makes it a brown dwarf, or "failed star" in the fuzzy border between planet and star, lacking the internal temperature and pressure to drive hydrogen fusion. The body's transits make it among the least massive for which an exact radius has been determined.

"I'm blown away by the scientific impact of the Kepler mission," Johnson said via e-mail. "It's not often that astronomers get order-of-magnitude gains in precision or sample size, but Kepler has provided both . . . My students and I are like kids in a candy store!"

And passionate hobbyist Apps says, "For an amateur to be involved . . . and to be published, that is as good as it gets, really." (Apps' successes have led a filmmaker to plan a movie dramatization of his adventures.)

RINGING LIKE BELLS

From the start, the Kepler mission recruited hundreds of specialists in the field of asteroseismology to help with the planet search. These scientists look at the flickers of stars to collect key data for deducing the size of an orbiting planet. But the scientists have been astonished at how much else they can tell about stars with Kepler's data. Don Kurtz, an astronomer at the University of Central Lancashire in Preston, England, is coauthor of what he calls the first asteroseismology textbook, published in 2009. Its foreword declares that the pulsations in a star's light due to sloshings of the star's atmosphere are physically similar to low-frequency sound waves.

Thus modern asteroseismology echoes a theory first put forward by Pythagoras that heavenly spheres make divine music. The vibrations of stars offer details about their internal structures, like a symphony's sound reveals the composition of the orchestra. The introduction to Kurtz's textbook says that just as "you can 'hear' the shape of the instrument, we can use the frequencies, amplitudes and phases of the sound waves that we detect in the stars to 'see' their interiors—to see their internal 'shapes.'"

Kurtz likened the technique to ultrasound exams that tell doctors the health and sex of a developing fetus. In fact, a team reported in the April 8 *Science* measurements from Kepler that reveal pulsations that echo all the way from the center of a star, a red giant.

"Until Kepler, I was doing the best stellar photometry in the business. The very best I could ever do, Kepler is doing it at least 10 or 20 times better, and most of the time it is 100 to 1,000 times better, while looking at 156,000 stars at once."

Kurtz recently examined Kepler's data on one star, known as KIC 10195926. It does something he has never seen: It quivers, or pulsates, on two axes of symmetry at once. The star is roughly twice the sun's mass and has now been classed as an Ap star, for A-peculiar, with a strong magnetic field. A paper by Kurtz and collaborators published online in April in the *Monthly Notices of the Royal Astronomical Society* reported that the star exhibits "torsional modes" in its rotation. Nothing like that has been seen before, either. It means the star's northern and southern halves trade off, one spinning faster than the other and then vice versa—a little bit like a dancer turning and doing the twist at the same time, shoulders and hips moving in opposite sync.

As such discoveries pile up, competition for time on ground-based telescopes in the Northern Hemisphere is already spiking for late summer, the peak of "Kepler Season" when the satellite's target is high in the Earth's night sky. Astronomers want to gather more information on objects that Kepler's data reveal to be interesting, or to make a case for objects that Kepler's handlers should add to their list.

A FURIOUS TANGO

Many of the stars Kepler is looking at have been visible in telescopes for centuries but, like most nondescript-looking stars, got little notice. What standard star catalogs say about seldom-examined stars is often sketchy or wrong. An example is HD 187091, about 1,000 light-years away. The Henry Draper Catalog of 1918–24 lists it as an ordinary A-class star, a type about twice as large and as massive as the sun. It did not seem unusual in any way.

But within just 10 days of operation, Kepler returned an astonishing light curve. Every 42 days the star's brightness rose to a sharp peak and quickly fell, regular as a metronome. The rise was less than 1 percent, "but to Kepler, that is huge," says William Welsh, an astronomy professor at San Diego State University. Furthermore, a complex forest of secondary brightness variations continued at a lower level in between the dramatic brightness peaks.

One hypothesis was that the star is orbited by a black hole, its strong gravity lensing the star's light into a beam that briefly aims at Earth once every orbit. A platoon of astronomers got time on NASA's Swift telescope to seek signs of a black hole's telltale X-rays, but no luck there. A spectroscopic examination with half a dozen ground-based telescopes solved the puzzle.

HD 187091, now known as KOI-54 for "Kepler Object of Interest No. 54," is not one A-star but two of nearly the same size, one 2.19 and the other 2.33 times the sun's 1.39-million-kilometer diameter. They are in a wildly stretched-out, elliptical orbit. Every 41 day, 19 hour orbit sends them racing nearly toward each other and zipping around one another at a separation equal to only about three times each one's diameter, and then flings them almost 120 million kilometers apart.

The brightening occurs as the stars, tidally warped by their gravity at closest approach into slight egg shapes, roast one another on their facing sides and heat up. And that explains the spike in brightness, the team reported online in February at arXiv.org. The more surprising revelation of Kepler's data is that one, and perhaps both, pulsate furiously at rates that are precise multiples of their rate of close encounters, in some cases pulsing exactly 90 and 91 times for each orbit. "Nobody had ever seen, or even thought, something like this could happen," Welsh says. Discovering that a star's rapid pulsations are not always driven by internal processes, but can be paced by a tidal metronome from a partner star, offers a new window into stellar dynamics and structure.

STAY TUNED

In the next year or two, and longer if the Kepler mission is extended, myriad more discoveries like these are likely. Eventually the craft will go out of commission; at the outside it could keep working for four or five years, says deputy Kepler project scientist Steve Howell, also at Ames.

But the telescope's abilities may be surpassed by another in the works. The European Space Agency is considering a mission called Plato, for Planetary Transits and Oscillations of stars. It could go up sometime between 2015 and 2025 with a prime mission much like Kepler's—to study stars with planetary systems. The craft is to be parked in a libration point, a sort of kink in the sun's and Earth's combined gravity that will hold it hovering near Earth but far enough away to be able to stare constantly for long periods at one region.

But unlike Kepler, Plato will be able to swivel around and look at different parts of the sky, and will use bundles of telescopes to stare at larger pieces with even greater precision. Undoubtedly it, too, will find yet more stars whose wonders have nothing to do with planets. Who knows: It may even be able to check on some that Kepler found first, just to see how they're doing.

Steven Weinberg

Symmetry: A "Key to Nature's Secrets"

FROM THE *NEW YORK REVIEW OF BOOKS*

The Nobel Prize–winning physicist Steven Weinberg considers symmetry: a useful scientific tool that, when examined closely, can lead to surprising conclusions about the adventitious nature of the universe—and our place in it.

WHEN I FIRST STARTED DOING RESEARCH IN THE late 1950s, physics seemed to me to be in a dismal state. There had been a great success a decade earlier in quantum electrodynamics, the theory of electrons and light and their interactions. Physicists then had learned how to calculate things like the strength of the electron's magnetic field with a precision unprecedented in all of science. But now we were confronted with newly discovered esoteric particles—muons and dozens of types of mesons and baryons—most existing nowhere in nature except in cosmic rays. And we had to deal with mysterious forces: strong nuclear forces that hold particles together inside atomic nuclei, and weak nuclear forces that can change the nature of these particles. We did not have a theory that would describe these particles and forces, and when we took a stab at a possible theory, we found that either we could not calculate its consequences, or when we could, we would come up with nonsensical results, like infinite energies or infinite probabilities. Nature, like an enemy, seemed intent on concealing from us its master plan.

At the same time, we did have a valuable key to nature's secrets. The laws of nature evidently obeyed certain principles of symmetry, whose consequences we could work out and compare with observation, even without a detailed theory of particles and forces. There were symmetries that dictated that certain distinct processes all go at the same rate, and that also dictated the existence of families of distinct particles that all have the same mass. Once we observed such equalities of rates or of masses, we could infer the existence of a symmetry, and this we thought would give us a clearer idea of the further observations that should be made, and of the sort of underlying theories that might or might not be possible. It was like having a spy in the enemy's high command.*

1.

I had better pause to say something about what physicists mean by principles of symmetry. In conversations with friends who are not physicists or mathematicians, I find that they often take symmetry to mean the identity of the two sides of something symmetrical, like the human face or a butterfly. That is indeed a kind of symmetry,

* This article is based in part on a talk given at a conference devoted to symmetry at the Technical University of Budapest in August 2009

but it is only one simple example of a huge variety of possible symmetries.

The Oxford English Dictionary tells us that symmetry is "the quality of being made up of exactly similar parts." A cube gives a good example. Every face, every edge, and every corner is just the same as every other face, edge, or corner. This is why cubes make good dice: if a cubical die is honestly made, when it is cast it has an equal chance of landing on any of its six faces.

The cube is one example of a small group of regular polyhedra—solid bodies with flat planes for faces, which satisfy the symmetry requirement that every face, every edge, and every corner should be precisely the same as every other face, edge, or corner. Thus the regular polyhedron called a triangular pyramid has four faces, each an equilateral triangle of the same size; six edges, at each of which two faces meet at the same angle; and four corners, at each of which three faces come together at the same angles.

These regular polyhedra fascinated Plato. He learned (probably from the mathematician Theaetetus) that regular polyhedra come in only five possible shapes, and he argued in *Timaeus* that these were the shapes of the bodies making up the elements: earth consists of little cubes, while fire, air, and water are made of polyhedra with four, eight, and twenty identical faces, respectively. The fifth regular polyhedron, with twelve identical faces, was supposed by Plato to symbolize the cosmos. Plato offered no evidence for all this—he wrote in *Timaeus* more as a poet than as a scientist, and the symmetries of these five bodies representing the elements evidently had a powerful hold on his poetic imagination.

The regular polyhedra in fact have nothing to do with the atoms that make up the material world, but they provide useful examples of a way of looking at symmetries, a way that is particularly congenial to physicists. A symmetry is a principle of invariance. That is, it tells us that something does not change its appearance when we make certain changes in our point of view—for instance, by rotating it or moving it. In addition to describing a cube by saying that it has six identical square faces, we can also say that its appearance does not change if we rotate it in certain ways—for instance by 90° around any direction parallel to the cube's edges.

The set of all such transformations of point of view that will leave a particular object looking the same is called that object's invariance group. This may seem like a fancy way of talking about things like cubes, but often in physics we make guesses about invariance groups, and test them experimentally, even when we know nothing else about the thing that is supposed to have the conjectured symmetry. There is a large and elegant branch of mathematics known as group theory, which catalogs and explores all possible invariance groups, and is described for general readers in two recently published books: *Symmetry: A Journey into the Patterns of Nature* by Marcus du Sautoy and *Why Beauty Is Truth: A History of Symmetry* by Ian Stewart.

2.

The symmetries that offered the way out of the problems of elementary particle physics in the 1950s were not the symmetries of objects, not even objects as important as atoms, but the symmetries of laws. A law of nature can be said to respect a certain symmetry if that law remains the same when we change the point of view from which we observe natural phenomena in certain definite ways. The particular set of ways that we can change our point of view without changing the law defines that symmetry.

Laws of nature, in the modern sense of mathematical equations that tell us precisely what will happen in various circumstances, first appeared as the laws of motion and gravitation that Newton developed as a basis for understanding Kepler's description of the solar system. From the beginning, Newton's laws incorporated symmetry: the laws that we observe to govern motion and gravitation do not change their form if we reset our clocks, or if we change the point from which distances are measured, or if we rotate our entire laboratory so it faces in a different direction.*

There is another less obvious symmetry, known today as Galilean invariance, that had been anticipated in the fourteenth century by Jean Buridan and Nicole Oresme: the laws of nature that we discover do not change their form if we observe nature within a moving laboratory, traveling at constant velocity. The fact that the earth is speeding around the sun, for instance, does not affect the laws of motion of material objects that we observe on the earth's surface.†

Newton and his successors took these principles of invariance pretty much for granted, as an implicit basis for their theories, so it was quite a wrench when these principles themselves became a subject of serious physical investigation. The crux of Einstein's 1905 Special Theory of Relativity was a modification of Galilean invariance. This was motivated in part by the persistent failure of physicists to find any effect of the earth's motion on the measured speed of light, analogous to the effect of a boat's motion on the observed speed of water waves.

It is still true in Special Relativity that making observations from a moving laboratory does not change the form of the observed laws of nature, but the effect of this motion on measured distances and times is different in Special Relativity from what Newton had thought. Motion causes lengths to shrink and clocks to slow down in such a way that the speed of light remains a constant, whatever the

---

* For reasons that are difficult to explain without mathematics, these symmetries imply important conservation laws: the conservation of energy, momentum, and angular momentum (or spin). Some other symmetries imply the conservation of other quantities, such as electric charge.

† Strictly speaking, Galilean invariance applies only approximately to the motion of the earth, since the earth is not moving in a straight line at constant speed. It is true that the earth's motion in its orbit does not affect the laws we observe, but this is because gravity balances the effects of the centrifugal force caused by the earth's curved motion. This too is dictated by a symmetry, but the symmetry here is Einstein's principle of general covariance, the basis of the general theory of relativity.

speed of the observer. This new symmetry, known as Lorentz invariance,* required profound departures from Newtonian physics, including the convertibility of energy and mass.

The advent and success of Special Relativity alerted physicists in the twentieth century to the importance of symmetry principles. But by themselves, the symmetries of space and time that are incorporated in the Special Theory of Relativity could not take us very far. One can imagine a great variety of theories of particles and forces that would be consistent with these space-time symmetries. Fortunately it was already clear in the 1950s that the laws of nature, whatever they are, also respect symmetries of other kinds, having nothing directly to do with space and time.

There are four forces that allow particles to interact with one another: the familiar gravity and electromagnetism, and the less well-known weak nuclear force (which is responsible for certain types of radioactive decay) and strong nuclear force (which binds protons and neutrons in the nucleus of an atom). (I am writing of a time, during the 1950s, before the formulation of the modern Standard Model, in which the three known forces other than gravity are now united in a single theory.) It had been known since the 1930s that the unknown laws that govern the strong nuclear force respect a symmetry between protons and neutrons, the two particles that make up atomic nuclei.

Even though the equations governing the strong forces were not known, the observations of nuclear properties had revealed that whatever these equations are, they must not change if everywhere in these equations we replace the symbol representing protons with that representing neutrons, and vice versa. Not only that, but the equations are also unchanged if we replace the symbols representing

---

* Lorentz had tried to explain the constancy of the observed speed of light by studying the effect of motion on particles of matter. Einstein was instead explaining the same observation by a change in one of nature's fundamental symmetries.

protons and neutrons with algebraic combinations of these symbols that represent superpositions of protons and neutrons, superpositions that might for instance have a 40 percent chance of being a proton and a 60 percent chance of being a neutron. It is like replacing a photo of Alice or of Bob with a picture in which photos of both Alice and Bob are superimposed. One consequence of this symmetry is that the nuclear force between two protons is not only equal to the force between two neutrons—it is also related to the force between a proton and a neutron.

Then as more and more types of particles were discovered, it was found in the 1960s that this proton–neutron symmetry was part of a larger symmetry group: not only are the proton and neutron related by this symmetry to each other, they are also related to six other subatomic particles, known as hyperons. The symmetry among these eight particles came to be called "the eightfold way." All the particles that feel the strong nuclear force fall into similar symmetrical families, with eight, ten, or more members.

But there was something puzzling about these internal symmetries: unlike the symmetries of space and time, these new symmetries were clearly neither universal nor exact. Electromagnetic phenomena did not respect these symmetries: protons and some hyperons are electrically charged; neutrons and other hyperons are not. Also, the masses of protons and neutrons differ by about 0.14 percent, and their masses differ from those of the lightest hyperon by 19 percent. If symmetry principles are an expression of the simplicity of nature at the deepest level, what are we to make of a symmetry that applies to only some forces, and even there is only approximate?

An even more puzzling discovery about symmetry was made in 1956–1957. The principle of mirror symmetry states that physical laws do not change if we observe nature in a mirror, which reverses distances perpendicular to the mirror (that is, something far behind your head looks in the mirror as if it is far behind your image, and hence far in front of you). This is not a rotation—there is no way of rotating your point of view that has the effect of reversing directions in and out of a mirror, but not sideways or vertically. It had generally been taken for granted that mirror symmetry, like the other symmetries of space and time, was exact and universal, but in 1957 experiments showed convincingly that, while the electromagnetic and strong nuclear forces do obey mirror symmetry, the weak nuclear force does not. Experiments showed, for example, that it was possible to distinguish a cobalt nucleus in the process of decaying—as a result of the weak nuclear force—from its mirror image, spinning in the opposite direction.

So we had a double mystery: What causes the observed violations of the eightfold way symmetry and of mirror symmetry? Theorists offered several possible answers, but as we will see, this was the wrong question.

The 1960s and 1970s witnessed a great expansion of our conception of the sort of symmetry that might be possible in physics. The approximate proton–neutron symmetry was originally understood to be rigid, in the sense that the equations governing the strong nuclear forces were supposed to be unchanged only if we changed protons and neutrons into mixtures of each other in the same way everywhere in space and time (physicists somewhat confusingly use the adjective "global" for what I am here calling rigid symmetries).

But what if the equations obeyed a more demanding symmetry, one that was local, in the sense that the equations would also be unchanged if we changed neutrons and protons into different mixtures of each other at different times and locations? In order to allow the different local mixtures to interact with one another without changing the equations, such a local symmetry would require some way for protons and neutrons to exert force on each other. Much as photons (the massless particles of light) are required to carry the electromagnetic force, a new massless particle, the gluon, would be needed to carry the force between protons and neutrons. It was hoped that this sort of theory of symmetrical forces might somehow explain the strong nuclear force that holds neutrons and protons together in atomic nuclei.

Conceptions of symmetry also expanded in a different direction. Theorists began in the 1960s to consider the possibility of symmetries that are "broken." That is, the underlying equations of physics might respect symmetries that are nevertheless not apparent in the actual physical states observed. The physical states that are possible in nature are represented by solutions of the equations of physics. When we have a broken symmetry, the solutions of the equations do not respect the symmetries of the equations themselves.*

The elliptical orbits of planets in the solar system provide a good example. The equations governing the gravitational field of the sun, and the motions of bodies in that field, respect rotational symmetry—there is nothing in these equations that distinguishes one direction in space from another. A circular planetary orbit of the sort imagined by Plato would also respect this symmetry, but the elliptical orbits actually encountered in the solar system do not: the long axis of an ellipse points in a definite direction in space.

At first it was widely thought that broken symmetry might have something to do with the small known violations of symmetries like mirror symmetry or the eightfold way. This was a false lead. A broken symmetry is nothing like an approximate symmetry, and is useless for putting particles into families like those of the eightfold way.

But broken symmetries do have consequences that can be checked empirically. Because of the spherical symmetry of the equations governing the sun's gravitational field, the long axis of an elliptical planetary orbit can point in any direction in space. This makes these

---

* Consider the equation x^3 equals x. This equation has a symmetry under the transformation that replaces x with $-x$; if we replace x with $-x$, we get the same equation. The equation has a solution $x = 0$, which respects the symmetry; $-0 = 0$. But it also has a solution in which $x = 1$. This does not respect the symmetry; -1 is not equal to 1. This is a broken symmetry. Of course, this equation is not much like the equations of physics.

orbits acutely sensitive to any small perturbation that violates the symmetry, like the gravitational field of other planets. For instance, these perturbations cause the long axis of Mercury's orbit to swing around 360° every 2,254 centuries.

In the 1960s theorists realized that the strong nuclear forces have a broken symmetry, known as chiral symmetry. Chiral symmetry is like the proton–neutron symmetry mentioned above, except that the symmetry transformations can be different for particles spinning clockwise or counterclockwise. The breaking of this symmetry requires the existence of the subatomic particles called pi mesons. The pi meson is in a sense the analog of the slow change in orientation of an elliptical planetary orbit; just as small perturbations can make large changes in an orbit's orientation, pi mesons can be created in collisions of neutrons and protons with relatively low energy.

The path out of the dismal state of particle physics in the 1950s turned out to lead through local and broken symmetries. First, electromagnetic and weak nuclear forces were found to be governed by a broken local symmetry. (The experiments now under way at Fermilab in Illinois and the new accelerator at CERN in Switzerland have as their first aim to pin down just what it is that breaks this symmetry.) Then the strong nuclear forces were found to be described by a different local symmetry. The resulting theory of strong, weak, and electromagnetic forces is what is now known as the Standard Model, and does a good job of accounting for virtually all phenomena observed in our laboratories.

3.

It would take far more space than I have here to go into details about these symmetries and the Standard Model, or about other proposed symmetries that go beyond those of the Standard Model. Instead I want to take up one aspect of symmetry that as far as I know has not yet been described for general readers. When the Standard Model was put in its present form in the early 1970s, theorists to their delight encountered something quite unexpected. It turned out that the Standard Model obeys certain symmetries that are accidental, in the sense that, though they are not the exact local symmetries on which the Standard Model is based, they are automatic consequences of the Standard Model. These accidental symmetries accounted for a good deal of what had seemed so mysterious in earlier years, and raised interesting new possibilities.

The origin of accidental symmetries lies in the fact that acceptable theories of elementary particles tend to be of a particularly simple type. The reason has to do with avoidance of the nonsensical infinities I mentioned at the outset. In theories that are sufficiently simple these infinities can be canceled by a mathematical process called "renormalization." In this process, certain physical constants, like masses and charges, are carefully redefined so that the infinite terms are canceled out, without affecting the results of the theory. In these simple theories, known as "renormalizable" theories, only a small number of particles can interact at any given location and time, and then the energy of interaction can depend in only a simple way on how the particles are moving and spinning.

For a long time many of us thought that to avoid intractable infinities, these renormalizable theories were the only ones physically possible. This posed a serious problem, because Einstein's successful theory of gravitation, the General Theory of Relativity, is not a renormalizable theory; the fundamental symmetry of the theory, known as general covariance (which says that the equations have the same form whatever coordinates we use to describe events in space and time), does not allow any sufficiently simple interactions. In the 1970s it became clear that there are circumstances in which nonrenormalizable theories are allowed without incurring nonsensical infinities, but that the relatively complicated interactions that make these theories nonrenormalizable are expected, under normal circumstances, to be so weak that physicists can usually ignore them and still get reliable approximate results.

This is a good thing. It means that to a good approximation there are only a few kinds of renormalizable theories that we need to consider as possible descriptions of nature.

Now, it just so happens that under the constraints imposed by Lorentz invariance and the exact local symmetries of the Standard Model, the most general renormalizable theory of strong and electromagnetic forces simply can't be complicated enough to violate mirror symmetry.* Thus, the mirror symmetry of the electromagnetic and strong nuclear forces is an accident, having nothing to do with any symmetry built into nature at a fundamental level. The weak nuclear forces do not respect mirror symmetry because there was never any reason why they should. Instead of asking what breaks mirror symmetry, we should have been asking why there should be any mirror symmetry at all. And now we know. It is accidental.

The proton–neutron symmetry is explained in a similar way. The Standard Model does not actually refer to protons and neutrons, but to the particles of which they are composed, known as quarks and gluons.† The proton consists of two quarks of a type called "up" and one of a type called "down"; the neutron consists of two down quarks and an up quark. It just so happens that in the most general renormalizable theory of quarks and gluons satisfying the symmetries of the Standard Model, the only things that can violate the proton–neutron symmetry are the masses of the quarks. The up and down quark masses are not at all equal—the down quark is nearly twice as heavy as the up quark—because there is no reason why they should be equal. But these masses are both very small—most of the masses of the protons and neutrons come from the strong nuclear force, not from the quark masses. To the extent that quark masses can be neglected, then, we have an accidental approximate symme-

---

* Honesty compels me to admit that here I am gliding over some technical complications.

† These particles are not observed experimentally, not because they are too heavy to be produced (gluons are massless, and some quarks are quite light), but because the strong nuclear forces bind them together in composite states like protons and neutrons.

try between protons and neutrons. Chiral symmetry and the eight-fold way arise in the same accidental way.

So mirror symmetry and the proton–neutron symmetry and its generalizations are not fundamental at all, but just accidents, approximate consequences of deeper principles. To the extent that these symmetries were our spies in the high command of nature, we were exaggerating their importance, as also often happens with real spies.

The recognition of accidental symmetry not only resolved the old puzzle about approximate symmetries; it also opened up exciting new possibilities. It turned out that there are certain symmetries that could not be violated in any theory that has the same particles and the same exact local symmetries as the Standard Model and that is simple enough to be renormalizable.* If really valid, these symmetries, known as lepton and baryon conservation,† would dictate that neutrinos (particles that feel only the weak and gravitational forces) have no mass, and that protons and many atomic nuclei are absolutely stable. Now, on experimental grounds these symmetries had been known long before the advent of the Standard Model, and had generally been thought to be exactly valid. But if they are actually accidental symmetries of the Standard Model, like the accidental proton–neutron symmetry of the strong forces, then they too might be only approximate. As I mentioned earlier, we now understand that interactions that make the theory nonrenormalizable are

---

* Again, I admit to passing over some technical complications.

† Lepton number is defined as the number of electrons and similar heavier charged particles plus the number of neutrinos, minus the number of their antiparticles. (This conservation law requires the neutrino to be massless because neutrinos and antineutrinos, respectively, spin only counterclockwise and clockwise around their directions of motion. If neutrinos have any mass then they travel at less than the speed of light, so it is possible to reverse their apparent direction of motion by travelling faster past them, hence converting the spin from counterclockwise to clockwise, and neutrinos to antineutrinos, which changes the lepton number.) Baryon number is proportional to the number of quarks minus the number of antiquarks.

not impossible, though they are likely to be extremely weak. Once one admits such more complicated nonrenormalizable interactions, the neutrino no longer has to be strictly massless, and the proton no longer has to be absolutely stable.

There are in fact possible nonrenormalizable interactions that would give the neutrino a tiny mass, of the order of one hundred millionth of the electron mass, and give protons a finite average lifetime, though one so long that typical protons in matter today will last much longer than the universe already has. Experiments in recent years have revealed that neutrinos do indeed have such masses. Experiments are under way to detect the tiny fraction of protons that decay in a year or so, and I would bet that these decays will eventually be observed. If protons do decay, the universe will eventually contain only lighter particles like neutrinos and photons. Matter as we know it will be gone.

I said that I would be concerned here with the symmetries of laws, not of objects, but there is one thing that is so important that I need to say a bit about it. It is the universe. As far as we can see, when averaged over sufficiently large scales containing many galaxies, the universe seems to have no preferred position, and no preferred directions—it is symmetrical. But this too may be an accident.

There is an attractive theory, called chaotic inflation, according to which the universe began without any special spatial symmetries, in a completely chaotic state. Here and there by accident the fields pervading the universe were more or less uniform, and according to the gravitational field equations it is these patches of space that then underwent an exponentially rapid expansion, known as inflation, leading to something like our present universe, with all nonuniformities in these patches smoothed out by the expansion. In different patches of space the symmetries of the laws of nature would be broken in different ways. Much of the universe is still chaotic, and it is only in the patches that inflated sufficiently (and in which symmetries were broken in the right ways) that life could arise, so any beings who study the universe will find themselves in such patches.

This is all quite speculative. There is observational evidence for an exponential early expansion, which has left its traces in the microwave radiation filling the universe, but as yet no evidence for an earlier period of chaos. If it turns out that chaotic inflation is correct, then much of what we observe in nature will be due to the accident of our particular location, an accident that can never be explained, except by the fact that it is only in such locations that anyone could live.

TIM FOLGER

Waiting for the Higgs

FROM *SCIENTIFIC AMERICAN*

With the Large Hadron Collider now online at CERN in Switzerland, physicists have their best chance of finding evidence for the existence of the elusive Higgs boson, the jewel in the crown of the Standard Model. But, as Tim Folger reports, some researchers at the recently shut down Tevatron in Illinois may have a chance of getting there first.

UNDERNEATH A RELICT PATCH OF ILLINOIS PRAIRIE, complete with a small herd of grazing buffalo, protons and antiprotons whiz along in opposite paths around a four-mile-long tunnel. And every second, hundreds of thousands of them slam together in a burst of obscure particles. It's another day at the Tevatron, a particle accelerator embedded in the verdant grounds of the

6,800-acre Fermi National Accelerator Laboratory complex in Bata-via, about 50 miles due west of Chicago. There have been many days like this one, some routine, some spectacular; of the 17 fundamental particles that physicists believe constitute all the ordinary matter and energy in the universe, three were discovered here. But there won't be many more such days. By October 1 the power supplies for more than 1,000 liquid-helium-cooled superconducting magnets will have been turned off forever, the last feeble stream of particles absorbed by a metal target, ending the 28-year run of what was until recently the most powerful particle accelerator in the world.

For several hundred physicists here who have spent nearly two decades searching for a hypothetical particle called the Higgs boson, the closure means ceding the hunt—and possible Nobel glory—to their archrival, the Large Hadron Collider, a newer, more powerful accelerator at CERN on the Swiss-French border. With its 17-mile circumference and higher energies, the LHC has displaced the Teva-tron as the world's premier particle physics research instrument, a position it will retain well into the next decade.

The U.S. Department of Energy's decision to shut down the Teva-tron at the close of this fiscal year did not surprise anyone at Fermi-lab. Some physicists had recommended that the DOE fund the aging accelerator for another three years, giving it a final crack at finding the elusive Higgs, a particle that theorists believe is responsible for endowing all other particles with mass. But even the most ardent Tevatron veterans admit that the old machine has finally been made redundant. "I don't have sadness," says Dmitri Denisov. "It's like your old car. The whole history of science is one of new tools. This one lasted for more than 25 years. It's time to move on."

That can't be an easy admission for Denisov, the co-spokesperson for the team that runs D-Zero, one of two hulking detectors that straddle the Tevatron. Two years ago, during a press conference at the annual meeting of the American Association for the Advancement of Science, Denisov said, "We now have a very, very good chance that we will see hints of the Higgs before the LHC will." At the time, an electrical failure had closed the LHC for several months, and Denisov's confidence was shared by many at Fermilab. But it was not to be. When the LHC came back online in November 2009, it quickly ramped up to energies three times higher than the Tevatron could match.

For the past three decades D-Zero's main competition has been the Tevatron's other enormous detector, the Collider Detector at Fermilab, or CDF, which sits atop the accelerator a grassy mile away from D-Zero. Hundreds of physicists from dozens of countries work at each.

This past spring physicists at the CDF announced that they had found hints in their data of what appeared to be a new particle. Might the Tevatron, in its waning days, have found the first telltale signs of the Higgs? Denisov and his colleagues at D-Zero immediately began to double-check the CDF results. As *Scientific American* went to press, the issue remained unsettled. Yet one thing is clear: the intra-accelerator competition is not yet over.

"I want to beat Dmitri, and vice versa," says Rob Roser, leader of the CDF team. "We're cordial; we talk; we're friends. But we always wanted to beat each other. Now the endgame is different. The LHC is the bad guy. It used to be Dmitri. I never wanted the LHC to beat either one of us. It's like, you can't beat up my little brother—only I can."

With old rivalries ending (almost) and new projects just starting, Fermilab is passing through an uncertain period. The same could be said for the entire discipline of particle physics. Physicists have been waiting a very long time for a machine that might give them access to some new realm of physical reality. Given that the LHC is expected to double its collision energies within the next two years, there is no shortage of ideas about what it might discover: extra dimensions, supersymmetry (the idea that every known particle has a so-called supersymmetric twin), the Higgs, of course. Best of all would be something completely unexpected. There is another possibility, however, usually dismissed but impossible to discount. And it simultaneously worries and intrigues physicists: What if the LHC, as well as the particle physics experiments planned at a Tevatron-less Fermilab for the next decade, finds nothing unexpected at all?

DESTINATION UNKNOWN

There was a time, not long ago, when physicists had many of the same hopes for the Tevatron that they now have for the LHC. Fifteen years before the LHC was turned on, physicists at Fermilab thought the Tevatron might bag the Higgs, find evidence for supersymmetry, identify the nature of dark matter, and more.

Besides netting a Nobel Prize, the discovery of the Higgs would provide the capstone to an illustrious era in physics. The Higgs boson is the last missing piece of the Standard Model, a complex theoretical edifice that describes the universe in terms of the interactions of the 17 fundamental particles. It unifies three of the four forces of nature: the strong force, which binds atomic nuclei; the weak force, which is responsible for particle decay; and the more familiar electromagnetic force. (Gravity is the only force not described by the Standard Model.) Theorists put the finishing touches on the Standard Model nearly 40 years ago, and since then every one of its predictions has been confirmed by experiment.

In 1995 the CDF and D-Zero teams made one of the most impressive confirmations with the discovery of the top quark—a massive elementary particle whose existence was first predicted in 1973. In that race, the Tevatron beat a European collider called the Super Proton Synchrotron, which is now used to feed particles into the LHC. It was the Tevatron's greatest triumph and established that the Standard Model was an incredibly accurate description of the universe, at least at the energies that physicists could probe with their best accelerators.

In 2001, after a five-year upgrade, the world's best accelerator became even better. Physicists hoped that the new, improved Tevatron would not only discover the Higgs—the last undiscovered piece of the Standard Model—but also uncover new phenomena lying beyond the Standard Model. For all the Standard Model's predictive power, physicists know that it cannot be a complete description of nature. Besides its failure to incorporate gravity, it has two other glaring shortcomings. The Standard Model provides no explanation of dark matter, which influences the motions of galaxies but otherwise does not seem to interact with ordinary matter. It also fails to account for dark energy, an utterly baffling phenomenon that appears to be accelerating the expansion of the universe.

But despite the upgrade, the Tevatron failed to move beyond the theory it had so spectacularly validated. "Ten years ago we anticipated cracking this nut, but we haven't yet," says Bob Tschirhart, a theoretical physicist at Fermilab. "There's a layer of existence out there that we haven't discovered. The Standard Model has been so good at predicting so much, but it has such obvious inadequacies. It's like an idiot savant."

In some sense, the legacy of the Tevatron is that the Standard Model works really, really well. It's no small achievement, but it was never intended to be the final goal. "We were supposed to find the Higgs, for sure," says Stephen Mrenna, a computational physicist who came to Fermilab in the mid-1990s. "And if super-symmetry was there, we were supposed to find it, too."

Physicists now hope that the LHC will succeed where the Tevatron failed by leading them into new territory and providing clues that might eventually enable them to replace the Standard Model. Mrenna, like most of his colleagues, believes that the LHC will find the Higgs sooner rather than later. "I think it will happen this year or next. That's where I would place my bet," he says. "If we don't find it, my belief that we won't find anything will go up greatly."

This is the problem with exploration: perhaps nothing is out there. Some physicists speculate that an "energy desert" exists between the realms they are able to probe now and the realm where truly new physics might emerge. If that's the case, new discoveries might be decades away. The LHC might be the most powerful accelerator ever built, but it is not so powerful that physicists can be completely sure it will punch through to another level of reality.

The real tool for that job was the Superconducting Super Collider (SSC), a machine that, at 54 miles in circumference, would have dwarfed the LHC. It would have been capable of generating particle beams with nearly three times the LHC's maximum energy. But cost overruns caused Congress to cancel the project in 1993, even though construction had already started near the small town of Waxahachie, Tex. "The SSC was designed from the beginning so that it would probe an energy scale where our expectations were that something new absolutely, positively had to happen," Mrenna says. "It really was the right collider to have built. The LHC is a cheap cousin. But it's good enough for now."

Unless, of course, it is not. If the LHC fails to find the Higgs or to make some other significant discovery, Mrenna says, it would become difficult for physicists to justify the costs of a more advanced accelerator. "You can ask what finding the Higgs boson has to do with the U.S. economy or the war on terror, or whatever," he observes, "and right now we get by saying the knowledge benefits everybody. People want to know how the universe works. And we're training lots of people, and it's always a good idea to take the cleverest people around and give them a really hard problem because usually there's a derivative that comes from it. But at some point the physics becomes less and less relevant."

In other words, if the energy desert is real, we may not be able to summon the will to cross it. "I'm actually a hanger-on from the SSC," Mrenna says. "I was a postdoc in its last year. And I have been waiting for a replacement for it ever since then, surviving in a rather grim job market. We need a success. We need to find something new."

NEXT LIFE

The world's first particle accelerator was made in 1929 by Ernest Lawrence, a physicist at the University of California, Berkeley. He called it a proton merry-go-round. It measured five inches across, was made of bronze, sealing wax and glass, and likely cost about $25. The LHC, which fired up about 80 years later, cost $10 billion. Its construction required an international effort, and it covers an area the size of a small town. Even if the LHC is wildly successful, there is little chance for a similar leap in scale in the foreseeable future.

"We know how to go 10 times higher in energy, but it would cost 10 times more," says Pier Oddone, director of Fermilab. "And we're already at the limit of what countries are willing to spend."

For the next decade and beyond the premier physics facility in the U.S. will live in the shadow of the LHC. Oddone says Fermilab will pursue a variety of projects that might have been delayed or canceled had the Tevatron remained in operation, but it is clear that the center of mass in the world of particle physics has shifted. "In an ideal world, we would have kept the Tevatron running without shutting down other stuff," he says. "But the money wasn't there." Experiments are now under way at Fermilab that will study the physics of neutrinos—probably the least understood of all fundamental particles—by shooting them from a source at Fermilab through 450 miles of the earth's crust toward a detector in a mine shaft in Minnesota. Fermilab scientists will also take part in the Dark Energy Survey, an astronomical investigation into the nature of dark energy.

But the overriding institutional goal is to once again host the world's most powerful particle accelerator. By 2020 Oddone hopes the lab will have completed construction of an accelerator called Project X. The near-term purpose of the mile-long machine will be to generate neutrinos and other particles for experiments at Fermilab. In the long term, the relatively small accelerator will serve as a test bed for technologies that might one day make it possible to build an affordable successor to the LHC.

"Project X is a bridge to getting back to the high-energy frontier of physics," says Steve Holmes, the project manager. "It's an opportunity to grab the leadership position and hold it. When people at lunch ask me what's the future for us here, I say that the U.S. led the world in high-energy physics for 70 years. It's the most fundamental field of physics, and as a great country we have to aspire to do that. What I can't tell them is when we'll get there."

We may not have heard the last from the Tevatron itself. Denisov, Roser and their colleagues at the old accelerator's two detectors have collected enough data to keep them busy for at least two years after the shutdown. The huge store of data could help flesh out initial discoveries made by the LHC. There is even an outside chance that some new result lies buried on a hard drive somewhere at Fermilab, just waiting to be analyzed. For a little while this past spring, it looked as if the Tevatron might have given us the first hint of physics beyond the Standard Model.

In April, Roser's CDF team announced that it had found very tentative evidence for a new particle or force of nature in data collected by the CDF. In a small but statistically significant number of cases, the physicists found a bump in the data, an excess of particles above what the Standard Model predicted. The particles appeared to be the decay products of some more massive particle, perhaps an unexpected form of the Higgs boson.

By the end of May the CDF team had analyzed the data again. "The bump is still there," Roser said at the time. Less than two weeks later, though, Roser's longtime colleague and rival Denisov said that the D-Zero team had completed an independent analysis of the CDF data. "We saw nothing," he said at a press conference.

It is not yet clear whether the bump will survive further scrutiny. The two groups are now comparing their results to see where the CDF analysis may have erred—if indeed it did err. For now, it looks like a new era in physics is on hold, as it has been for more than 30 years. It will be a shame if the bump vanishes. Discovering the Higgs would have made for quite an exit for the Tevatron. Within the next year or so we might all find out if the LHC can do any better.

DEVIN POWELL

Moved by Light

FROM *SCIENCE NEWS*

The weirdness of the quantum world is well documented—subatomic particles can exist in two states at once and seem to defy classical notions of cause and effect. Devin Powell investigates a group of experiments in which macroscopic objects, supercooled by beams of light, can be teased into demonstrating this kind of quantum behavior.

WELCOME TO QUANTUMVILLE. POPULATION: UNcertain. Walk down Main Street, lined with blurry cars simultaneously moving and remaining still. See the house with the curtains drawn? The television in the living room is both on and off at the same time. In this neighborhood, everyday objects do seemingly contradictory things.

You won't drive through this farfetched town anytime soon, but it's not as far off the map as it used to be. In laboratories across the world, bits of metal and glass are being groomed to behave in ways that defy common sense. Objects big enough to be seen and touched—some weighing kilograms—are beginning to rebel against the physical laws that govern daily experience.

At the forefront of this effort is a growing discipline called optomechanics. Its practitioners use beams of light to do something utterly unfeasible a decade ago: make large objects colder than they would be in the void of outer space. Only at these temperatures do objects reach energies low enough to enter the realm of quantum mechanics and start behaving like subatomic particles.

"Our guiding principle is to see quantum effects in a macroscopic object," says physicist Ray Simmonds of the National Institute of Standards and Technology in Boulder, Colo.

A number of optomechanics teams have sprung up in recent years, each cooling its own favorite bit of fairly ordinary stuff. Simmonds works with an aluminum drum (unveiled in the March 10 *Nature*). In Switzerland, scientists chill silica doughnuts. At Yale University, saillike membranes are the vogue.

"We're putting the mechanics back in quantum mechanics," says Yale physicist Jack Harris.

It's mainly a race of tortoises creeping steadily closer to absolute zero, the coldest of the cold. But recently an interloper hare took a shortcut to the lead. And the stakes are high: The winners will test whether quantum mechanics holds at ever-larger scales and may go on to build a new generation of mechanical devices useful in quantum computing.

COOLING TOUCH

Spend an afternoon watching sunbathers burn at the beach, and the idea of using light to refrigerate may seem counterintuitive. But light particles have a hidden cooling ability that comes from the tiny nudge they impart when bouncing off an object. This force, too weak for a beachgoer to feel, is so feeble that sunlight reflecting off a square-meter mirror delivers a pressure less than a thousandth of the weight of a small paper clip. "It's an incredibly tiny effect," says physicist Steve Girvin, also of Yale.

In the 1970s scientists figured out how to use this "radiation pressure" to cool individual atoms by damping their vibrations with lasers. Now a slew of new devices leverage the punch of light and other forms of electromagnetic energy to cool objects made of trillions of atoms or more. This scaled-up cooling doesn't suppress the vibrations of individual atoms. Instead, it quiets the inherent wobbling of an entire object, like a foot pressed to a flopping diving board.

Putting light's cooling power to work starts with a laser beam bouncing between two mirrors. The distance between the mirrors in this "optical cavity" determines the frequency of light that will resonate—just as the length of a guitar string determines its pitch. Keep the mirrors still and properly tuned light will bounce back and forth, as constant as a metronome.

But allow one of these mirrors to wobble, and a more intricate and subtle interplay emerges. A laser beam tuned below the resonance frequency of the cavity will push against the swaying mirror and snatch away energy. By stealing vibrational energy from the mirror, the bouncing light gets a boost up to the optical cavity's stable frequency. Robbed of energy, the mirror's swaying weakens, and it cools.

By measuring the light leaking out of this type of system, two groups of physicists showed in 2006 that they could cool mirrors to 10 kelvins (10 degrees Celsius above absolute zero). A third used a similar technique to cool a glass doughnut to 11 kelvins, colder than the object would be if it were wobbling on the dark side of the moon.

"This demonstration that you could use laser radiation to cool a mechanical object, this started the race," says Tobias Kippenberg, leader of the doughnut team at the Swiss Federal Institute of Technology in Lausanne. "Every year we improve our cooling by a factor of 10."

As papers flowed in and objects neared the bottom of the thermometer, researchers competed to suck out every last drop of energy. The goal: to reach the ground state, where an object no longer possesses any packets, or quanta, of vibrational energy. In this state, motion almost completely stops and the quantum regime begins to become a reality.

But getting those last few quanta out would be a challenging task; even the mirrors at 10 kelvins still contained tens of thousands to hundreds of thousands of quanta.

Better lasers and equipment refinements allowed three groups, publishing in *Nature Physics* in 2009, to reach 63, 37 and 30 quanta. Keith Schwab of Caltech bombarded a wobbling object with microwaves that drained away all but about four quanta. He and his colleagues reported in *Nature* in 2010 that they had put their object into its ground state 21 percent of the time—tantalizingly close to the consistency needed to test for quantum effects.

Then in April 2010, a shot rang out. An object had been spotted entering its ground state over and over again—by an outsider who wasn't even using light.

"I wanted to get to the ground state in the quickest and most efficient way possible and have there be no question that I was there," says Andrew Cleland, a physicist at the University of California, Santa Barbara, who reported his team's achievement in *Nature* (SN 4/10/10, p. 10).

Cleland's secret: While other scientists built stuff that shook thousands or millions of times a second, he created a ceramic wafer 30 micrometers long that expanded and contracted 6 billion times per second. The faster an object's natural quiver, the easier it is to remove energy, meaning less cooling needed to reach the ground state. Using a state-of-the-art liquid-helium refrigerator capable of achieving millikelvin temperatures, Cleland's team put the wafer in its ground state 93 percent of the time.

By measuring the electric fields produced by this object, Cleland and his colleagues showed that they could nudge the wafer into a state of superposition—both moving and still at the same time.

"There can be no doubt that we achieved superposition," Cleland says. This first demonstration of quantum effects in a fairly ordinary object was named the 2010 Breakthrough of the Year by *Science*.

But Cleland's sprint to the front of the pack has some long-term disadvantages. His technique is blind to the actual position of a fluctuating object, for one thing, and thus he can't spot one of the consequences of quantum mechanics: zero-point energy, which gives an object residual motion even in its ground state. Experimentalists using optomechanics hope to detect this motion and verify that it is proportional to how fast an object normally wobbles.

BACK IN FRONT

Girding themselves for the long haul, optomechanics teams have now begun to catch up to Cleland's hare strategy. On March 21 in Dallas at the American Physical Society meeting, members of the NIST team presented data showing that their drumlike membrane had reached the ground state about 60 percent of the time.

The aluminum skin of this drum—in technical terms, a resonator—moves up and down much more slowly than Cleland's object, vibrating less than 11 million times per second. Reaching the ground state at this slower wobble couldn't be done with Cleland's refrigerator; it required the cooling nudge of microwaves.

The payoff for going the extra mile: time. The slower an object wobbles, the longer it tends to stay in its ground state. For Cleland, the ground state lifetime was about 6 nanoseconds. "The difference with our system, our resonator, is that it has a very long lifetime, about 100 microseconds," says Simmonds. "That's the key element that sets it apart."

With the results unpublished, the team won't say whether any quantum effects have been seen. But the stability could give the researchers an advantage for using optomechanical devices to store and relay information.

A "killer app," some say, would be playing interpreter between different wavelengths of light or other electromagnetic energy. A resonator in its ground state could theoretically be designed to absorb photons of just about any kind of light, stored as packets of vibrational energy.

Cool the resonator back to its ground state, and it could release this energy as light of a different wavelength. So gigahertz microwave energy that sets a stick to wobbling could be reemitted at optical frequencies hundreds of thousands of times higher, for instance. Such devices could bridge quantum computing systems that use different frequencies of light to transmit bits of information.

At Caltech, applied physicist Oskar Painter is taking steps toward realizing this light-to-light conversion at higher temperatures. He designs nanometer-scale optomechanical crystals that convert higher-frequency light to lower-frequency vibrations. A zipperlike object described in 2009 in *Nature*, for instance, could one day be useful for converting optical light into microwaves.

Optomechanical techniques, such as those used by Painter, could also shave the sensitivities of force detectors. At Yale, engineer Hong Tang develops sensors out of light-cooled resonators that promise unprecedentedly low levels of background noise.

"We want to make better accelerometers and better inertia sensors," Tang says. These devices, similar to those that sense the motion of a Wii controller, could measure tiny changes in movement and direction.

Like many other optomechanics researchers, Painter and Tang receive funding from the Defense Advanced Research Projects Agency. DARPA hopes to use laser-cooled sensors to improve the ability of vehicles to navigate underwater, says DARPA program manager Jamil Abo-Shaeer. "We want to push these things to the limits of quantum mechanics, the ultimate limit," he says.

While DARPA funds the development of devices that can't even be seen without a microscope, other scientists are putting optomechanics to work cooling some of the largest detectors in the world: the gravitational wave detectors of the LIGO project, built to search for gentle ripples in spacetime thought to be produced by (among other cosmic events) colliding black holes.

Chasing ever greater sensitivities, these researchers use lasers to still the vibrations of their detectors' giant mirrors—the behemoths of the optomechanical world, weighing in at more than 10 kilograms. Despite their immense size, these mirrors have now been cooled to 234 quanta, MIT quantum physicist Nergis Mavalvala and LIGO colleagues reported in 2009 in the *New Journal of Physics*. "Our challenges are really the same as everyone else's, but we need to somehow cool our gram and kilogram-sized objects to nanokelvins," says Mavalvala.

Working on another gravitational wave detector called AURIGA, researchers in Italy set the record for largest object effectively cooled via optomechanics. An aluminum bar weighing more than 1 ton reached a mere 4,000 quanta, the team reported in *Physical Review Letters* in 2008.

Whether such large mirrors and bars could ever demonstrate quantum effects, though, is an open question. In principle, some physicists say, quantum mechanics should hold for objects of any size. "We don't know of any fundamental limit," Harris says.

Practical considerations may ultimately limit the size of quantum objects, though. Any observation, be it by a pair of eyes or a stray, colliding air molecule, can destroy a quantum state. The larger an object is, the harder it is to keep isolated. But that isn't stopping researchers with bigger objects from lining up behind Cleland and the NIST team to stretch the bounds on quantum effects.

"If we can prove that quantum mechanics holds for larger and larger objects, that would be quite spectacular," says Dirk Bouw-meester of UC Santa Barbara. "But it would also be spectacular if we can prove that it doesn't. New theories would be needed."

One of the slowest tortoises in the race, Bouwmeester's pace is deliberate. His mirrors, tens of micrometers across, vibrate a mere 10,000 or so times per second and promise an extended quantum lifetime. This durability, he says, is needed to test a controversial idea that gravity and quantum weirdness can't coexist for long at everyday scales.

More than three-quarters of a century of research has made scientists more comfortable with quantum mechanics at small scales, but supersizing it can seem as bizarre today as it did to Erwin Schrödinger. In 1935, he poked fun at the idea in his famous thought experiment: a cat in a box that could be both alive and dead at the same time, as long as no one peeked inside the box and forced a choice, killing with curiosity.

Perhaps it is still too much to imagine Schrödinger's cat behind the drawn curtains of Quantumville's homes, simultaneously nibbling Purina in three different rooms at once. But as researchers continue to cool knickknack after knickknack in their optomechanical grab bag, they may catch at least a faint echo of a meow.

ALAN LIGHTMAN

The Accidental Universe

FROM *HARPER'S MAGAZINE*

> *Many theoretical physicists now think that our universe is but one part of a larger "multiverse"—where other universes are conjectured although their existence cannot be truly proved. The physicist and novelist Alan Lightman asks whether that possibility makes physics more like theology than science.*

IN THE FIFTH CENTURY B.C., THE PHILOSOPHER DEMocritus proposed that all matter was made of tiny and indivisible atoms, which came in various sizes and textures—some hard and some soft, some smooth and some thorny. The atoms themselves were taken as givens. In the nineteenth century, scientists discovered that the chemical properties of atoms repeat periodically (and created the periodic table to reflect this fact), but the origins of such patterns remained mysterious. It wasn't until the twentieth century that scientists learned that the properties of an atom are determined by the number and placement of its electrons, the subatomic particles that orbit its nucleus. And we now know that all atoms heavier than helium were created in the nuclear furnaces of stars.

The history of science can be viewed as the recasting of phenomena that were once thought to be accidents as phenomena that can be understood in terms of fundamental causes and principles. One can add to the list of the fully explained: the hue of the sky, the orbits of planets, the angle of the wake of a boat moving through a lake, the six-sided patterns of snowflakes, the weight of a flying bustard, the temperature of boiling water, the size of raindrops, the circular shape of the sun. All these phenomena and many more, once thought to have been fixed at the beginning of time or to be the result of random events thereafter, have been explained as necessary consequences of the fundamental laws of nature—laws discovered by human beings.

This long and appealing trend may be coming to an end. Dramatic developments in cosmological findings and thought have led some of the world's premier physicists to propose that our universe is only one of an enormous number of universes with wildly varying properties, and that some of the most basic features of our particular universe are indeed mere accidents—a random throw of the cosmic dice. In which case, there is no hope of ever explaining our universe's features in terms of fundamental causes and principles.

It is perhaps impossible to say how far apart the different universes may be, or whether they exist simultaneously in time. Some may have stars and galaxies like ours. Some may not. Some may be finite in size. Some may be infinite. Physicists call the totality of universes the "multiverse." Alan Guth, a pioneer in cosmological thought, says that "the multiple-universe idea severely limits our hopes to understand the world from fundamental principles." And the philosophical ethos of science is torn from its roots. As put to me recently by Nobel Prize–winning physicist Steven Weinberg, a man as careful in his words as in his mathematical calculations, "We now find ourselves at a historic fork in the road we travel to understand the laws of nature. If the multiverse idea is correct, the style of fundamental physics will be radically changed."

The scientists most distressed by Weinberg's "fork in the road" are theoretical physicists. Theoretical physics is the deepest and purest branch of science. It is the outpost of science closest to philosophy, and religion. Experimental scientists occupy themselves with observing and measuring the cosmos, finding out what stuff exists, no matter how strange that stuff may be. Theoretical physicists, on the other hand, are not satisfied with observing the universe. They want to know why. They want to explain all the properties of the universe in terms of a few fundamental principles and parameters. These fundamental principles, in turn, lead to the "laws of nature," which govern the behavior of all matter and energy. An example of a fundamental principle in physics, first proposed by Galileo in 1632 and extended by Einstein in 1905, is the following: All observers traveling at constant velocity relative to one another should witness identical laws of nature. From this principle, Einstein derived his theory of special relativity. An example of a fundamental parameter is the mass of an electron, considered one of the two dozen or so "elementary" particles of nature. As far as physicists are concerned, the fewer the fundamental principles and parameters, the better. The underlying hope and belief of this enterprise has always been that these basic principles are so restrictive that only one, self-consistent universe is possible, like a crossword puzzle with only one solution. That one universe would be, of course, the universe we live in. Theoretical physicists are Platonists. Until the past few years, they agreed that the entire universe, the one universe, is generated from a few mathematical truths and principles of symmetry, perhaps throwing in a handful of parameters like the mass of the electron. It seemed that we were closing in on a vision of our universe in which everything could be calculated, predicted, and understood.

However, two theories in physics, eternal inflation and string theory, now suggest that the same fundamental principles from which the laws of nature derive may lead to many different self-consistent universes, with many different properties. It is as if you walked into a shoe store, had your feet measured, and found that a size 5 would fit you, a size 8 would also fit, and a size 12 would fit equally well. Such wishywashy results make theoretical physicists extremely unhappy. Evidently, the fundamental laws of nature do not pin down a single and unique universe. According to the current thinking of many physicists, we are living in one of a vast number of universes. We are living in an accidental universe. We are living in a universe uncalculable by science.

"BACK IN THE 1970S AND 1980S," says Alan Guth, "the feeling was that we were so smart, we almost had everything figured out." What physicists had figured out were very accurate theories of three of the four fundamental forces of nature: the strong nuclear force that binds atomic nuclei together, the weak force that is responsible for some forms of radioactive decay, and the electromagnetic force between electrically charged particles. And there were prospects for merging the theory known as quantum physics with Einstein's theory of the fourth force, gravity, and thus pulling all of them into the fold of what physicists called the Theory of Everything, or the Final Theory. These theories of the 1970s and 1980s required the specification of a couple dozen parameters corresponding to the masses of the elementary particles, and another half dozen or so parameters corresponding to the strengths of the fundamental forces. The next step would then have been to derive most of the elementary particle masses in terms of one or two fundamental masses and define the strengths of all the fundamental forces in terms of a single fundamental force.

There were good reasons to think that physicists were poised to take this next step. Indeed, since the time of Galileo, physics has been extremely successful in discovering principles and laws that have fewer and fewer free parameters and that are also in close agreement with the observed facts of the world. For example, the observed rotation of the ellipse of the orbit of Mercury, 0.012 degrees per century, was successfully calculated using the theory of general relativity, and the observed magnetic strength of an electron, 2.002319 magnetons, was derived using the theory of quantum electrodynamics. More than any other science, physics brims with highly accurate agreements between theory and experiment.

Guth started his physics career in this sunny scientific world. Now sixty-four years old and a professor at MIT, he was in his early thirties when he proposed a major revision to the Big Bang theory, something called inflation. We now have a great deal of evidence suggesting that our universe began as a nugget of extremely high density and temperature about 14 billion years ago and has been expanding, thinning out, and cooling ever since. The theory of inflation proposes that when our universe was only about a trillionth of a trillionth of a trillionth of a second old, a peculiar type of energy caused the cosmos to expand very rapidly. A tiny fraction of a second later, the universe returned to the more leisurely rate of expansion of the standard Big Bang model. Inflation solved a number of outstanding problems in cosmology, such as why the universe appears so homogeneous on large scales.

When I visited Guth in his third-floor office at MIT one cool day in May, I could barely see him above the stacks of paper and empty Diet Coke bottles on his desk. More piles of paper and dozens of magazines littered the floor. In fact, a few years ago Guth won a contest sponsored by the *Boston Globe* for the messiest office in the city. The prize was the services of a professional organizer for one day. "She was actually more a nuisance than a help. She took piles of envelopes from the floor and began sorting them according to size." He wears aviator-style eyeglasses, keeps his hair long, and chain-drinks

Diet Cokes. "The reason I went into theoretical physics," Guth tells me, "is that I liked the idea that we could understand everything—i.e., the universe—in terms of mathematics and logic." He gives a bitter laugh. We have been talking about the multiverse.

WHILE CHALLENGING THE PLATONIC DREAM of theoretical physicists, the multiverse idea does explain one aspect of our universe that has unsettled some scientists for years: according to various calculations, if the values of some of the fundamental parameters of our universe were a little larger or a little smaller, life could not have arisen. For example, if the nuclear force were a few percentage points stronger than it actually is, then all the hydrogen atoms in the infant universe would have fused with other hydrogen atoms to make helium, and there would be no hydrogen left. No hydrogen means no water. Although we are far from certain about what conditions are necessary for life, most biologists believe that water is necessary. On the other hand, if the nuclear force were substantially weaker than what it actually is, then the complex atoms needed for biology could not hold together. As another example, if the relationship between the strengths of the gravitational force and the electromagnetic force were not close to what it is, then the cosmos would not harbor any stars that explode and spew out life-supporting chemical elements into space or any other stars that form planets. Both kinds of stars are required for the emergence of life. The strengths of the basic forces and certain other fundamental parameters in our universe appear to be "fine-tuned" to allow the existence of life. The recognition of this fine-tuning led British physicist Brandon Carter to articulate what he called the anthropic principle, which states that the universe must have the parameters it does because we are here to observe it. Actually, the word anthropic, from the Greek for "man," is a misnomer: if these fundamental parameters were much different from what they are, it is not only human beings who would not exist. No life of any kind would exist.

If such conclusions are correct, the great question, of course, is why these fundamental parameters happen to lie within the range needed for life. Does the universe care about life? Intelligent design is one answer. Indeed, a fair number of theologians, philosophers, and even some scientists have used fine-tuning and the anthropic principle as evidence of the existence of God. For example, at the 2011 Christian Scholars' Conference at Pepperdine University, Francis Collins, a leading geneticist and director of the National Institutes of Health, said, "To get our universe, with all of its potential for complexities or any kind of potential for any kind of life-form, everything has to be precisely defined on this knife edge of improbability. . . . [Y]ou have to see the hands of a creator who set the parameters to be just so because the creator was interested in something a little more complicated than random particles."

Intelligent design, however, is an answer to fine-tuning that does not appeal to most scientists. The multiverse offers another explanation. If there are countless different universes with different properties—for example, some with nuclear forces much stronger than in our universe and some with nuclear forces much weaker—then some of those universes will allow the emergence of life and some will not. Some of those universes will be dead, lifeless hulks of matter and energy, and others will permit the emergence of cells, plants and animals, minds. From the huge range of possible universes predicted by the theories, the fraction of universes with life is undoubtedly small. But that doesn't matter. We live in one of the universes that permits life because otherwise we wouldn't be here to ask the question.

The explanation is similar to the explanation of why we happen to live on a planet that has so many nice things for our comfortable existence: oxygen, water, a temperature between the freezing and boiling points of water, and so on. Is this happy coincidence just good luck, or an act of Providence, or what? No, it is simply that we could not live on planets without such properties. Many other planets exist that are not so hospitable to life, such as Uranus, where the temperature is −371 degrees Fahrenheit, and Venus, where it rains sulfuric acid.

The multiverse offers an explanation to the fine-tuning conundrum that does not require the presence of a Designer. As Steven Weinberg says: "Over many centuries science has weakened the hold of religion, not by disproving the existence of God but by invalidating arguments for God based on what we observe in the natural world. The multiverse idea offers an explanation of why we find ourselves in a universe favorable to life that does not rely on the benevolence of a creator, and so if correct will leave still less support for religion."

Some physicists remain skeptical of the anthropic principle and the reliance on multiple universes to explain the values of the fundamental parameters of physics. Others, such as Weinberg and Guth, have reluctantly accepted the anthropic principle and the multiverse idea as together providing the best possible explanation for the observed facts.

If the multiverse idea is correct, then the historic mission of physics to explain all the properties of our universe in terms of fundamental principles—to explain why the properties of our universe must necessarily be what they are—is futile, a beautiful philosophical dream that simply isn't true. Our universe is what it is because we are here. The situation could be likened to a school of intelligent fish who one day began wondering why their world is completely filled with water. Many of the fish, the theorists, hope to prove that the entire cosmos necessarily has to be filled with water. For years, they put their minds to the task but can never quite seem to prove their assertion. Then, a wizened group of fish postulates that maybe they are fooling themselves. Maybe there are, they suggest, many other worlds, some of them completely dry, and everything in between.

THE MOST STRIKING EXAMPLE OF fine-tuning, and one that practically demands the multiverse to explain it, is the unexpected detection of what scientists call dark energy. Little more than a decade ago, using robotic telescopes in Arizona, Chile, Hawaii, and outer space that can comb through nearly a million galaxies a night, astronomers discovered that the expansion of the universe is accelerating. As mentioned previously, it has been known since the late 1920s that the universe is expanding; it's a central feature of the Big Bang model. Orthodox cosmological thought held that the expansion is slowing down. After all, gravity is an attractive force; it pulls masses closer together. So it was quite a surprise in 1998 when two teams of astronomers announced that some unknown force appears to be jamming its foot down on the cosmic accelerator pedal. The expansion is speeding up.

Galaxies are flying away from each other as if repelled by antigravity. Says Robert Kirshner, one of the team members who made the discovery: "This is not your father's universe." (In October, members of both teams were awarded the Nobel Prize in Physics.)

Physicists have named the energy associated with this cosmological force dark energy. No one knows what it is. Not only invisible, dark energy apparently hides out in empty space. Yet, based on our observations of the accelerating rate of expansion, dark energy constitutes a whopping three quarters of the total energy of the universe. It is the invisible elephant in the room of science.

The amount of dark energy, or more precisely the amount of dark energy in every cubic centimeter of space, has been calculated to be about one hundred-millionth (10^{-8}) of an erg per cubic centimeter. (For comparison, a penny dropped from waist-high hits the floor with an energy of about three hundred thousand—that is, 3×10^5— ergs.) This may not seem like much, but it adds up in the vast volumes of outer space. Astronomers were able to determine this number by measuring the rate of expansion of the universe at different epochs—if the universe is accelerating, then its rate of expansion was slower in the past. From the amount of acceleration, astronomers can calculate the amount of dark energy in the universe.

Theoretical physicists have several hypotheses about the identity of dark energy. It may be the energy of ghostly subatomic particles that can briefly appear out of nothing before self-annihilating and slipping back into the vacuum. According to quantum physics, empty space is a pandemonium of subatomic particles rushing about and then vanishing before they can be seen. Dark energy may also be associated with an as-yet-unobserved force field called the Higgs field, which is sometimes invoked to explain why certain kinds of matter have mass. (Theoretical physicists ponder things that other people do not.) And in the models proposed by string theory, dark energy may be associated with the way in which extra dimensions of space—beyond the usual length, width, and breadth—get compressed down to sizes much smaller than atoms, so that we do not notice them.

These various hypotheses give a fantastically large range for the theoretically possible amounts of dark energy in a universe, from something like 10^{115} ergs per cubic centimeter to -10^{115} ergs per cubic centimeter. (A negative value for dark energy would mean that it acts to decelerate the universe, in contrast to what is observed.) Thus, in absolute magnitude, the amount of dark energy actually present in our universe is either very, very small or very, very large compared with what it could be. This fact alone is surprising. If the theoretically possible positive values for dark energy were marked out on a ruler stretching from here to the sun, with zero at one end of the ruler and 10^{115} ergs per cubic centimeter at the other end, the value of dark energy actually found in our universe (10^{-8} ergs per cubic centimeter) would be closer to the zero end than the width of an atom.

On one thing most physicists agree: If the amount of dark energy in our universe were only a little bit different than what it actually is, then life could never have emerged. A little more and the universe would accelerate so rapidly that the matter in the young cosmos could never pull itself together to form stars and thence form the complex atoms made in stars. And, going into negative values of dark energy, a little less and the universe would decelerate so rapidly that it would recollapse before there was time to form even the simplest atoms.

Here we have a clear example of fine-tuning: out of all the possible amounts of dark energy that our universe might have, the actual amount lies in the tiny sliver of the range that allows life. There is little argument on this point. It does not depend on assumptions about whether we need liquid water for life or oxygen or particular biochemistries. As before, one is compelled to ask the question: Why does such fine-tuning occur? And the answer many physicists now believe: The multiverse. A vast number of universes may exist, with many different values of the amount of dark energy. Our particular universe is one of the universes with a small value, permitting the emergence of life. We are here, so our universe must be such a universe. We are an accident. From the cosmic lottery hat containing zillions of universes, we happened to draw a universe that allowed life. But then again, if we had not drawn such a ticket, we would not be here to ponder the odds.

THE CONCEPT OF THE MULTIVERSE is compelling not only because it explains the problem of fine-tuning. As I mentioned earlier, the possibility of the multiverse is actually predicted by modern theories of physics. One such theory, called eternal inflation, is a revision of Guth's inflation theory developed by Andrei Linde, Paul Steinhardt, and Alex Vilenkin in the early and mid-1980s. In regular inflation theory, the very rapid expansion of the infant universe is caused by an energy field, like dark energy, that is temporarily trapped in a condition that does not represent the lowest possible energy for the universe as a whole—like a marble sitting in a small dent on a table. The marble can stay there, but if it is jostled it will roll out of the dent, roll across the table, and then fall to the floor (which represents the lowest possible energy level). In the theory of eternal inflation, the dark energy field has many different values at different points of space, analogous to lots of marbles sitting in lots of dents on the cosmic table. Moreover, as space expands rapidly, the number of marbles increases. Each of these marbles is jostled by the random processes inherent in quantum mechanics, and some of the marbles will begin rolling across the table and onto the floor. Each marble starts a new Big Bang, essentially a new universe. Thus, the original, rapidly expanding universe spawns a multitude of new universes, in a never-ending process.

String theory, too, predicts the possibility of the multiverse. Originally conceived in the late 1960s as a theory of the strong nuclear force but soon enlarged far beyond that ambition, string theory postulates that the smallest constituents of matter are not subatomic particles like the electron but extremely tiny one-dimensional "strings" of energy. These elemental strings can vibrate at different frequencies, like the strings of a violin, and the different modes of vibration correspond to different fundamental particles and forces. String theories typically require seven dimensions of space in addition to the usual three, which are compacted down to such small sizes that we never experience them, like a three-dimensional garden hose that appears as a one-dimensional line when seen from a great distance. There are, in fact, a vast number of ways that the extra dimensions in string theory can be folded up, and each of the different ways corresponds to a different universe with different physical properties.

It was originally hoped that from a theory of these strings, with very few additional parameters, physicists would be able to explain all the forces and particles of nature—all of reality would be a manifestation of the vibrations of elemental strings. String theory would then be the ultimate realization of the Platonic ideal of a fully explicable cosmos.

In the past few years, however, physicists have discovered that string theory predicts not a unique universe but a huge number of possible universes with different properties. It has been estimated that the "string landscape" contains 10^{500} different possible universes. For all practical purposes, that number is infinite.

It is important to point out that neither eternal inflation nor string theory has anywhere near the experimental support of many previous theories in physics, such as special relativity or quantum electrodynamics, mentioned earlier. Eternal inflation or string theory, or both, could turn out to be wrong. However, some of the world's leading physicists have devoted their careers to the study of these two theories.

BACK TO THE INTELLIGENT FISH. The wizened old fish conjecture that there are many other worlds, some with dry land and some with water. Some of the fish grudgingly accept this explanation. Some feel relieved. Some feel like their lifelong ruminations have been pointless. And some remain deeply concerned. Because there is no way they can prove this conjecture. That same uncertainty disturbs many physicists who are adjusting to the idea of the multiverse. Not only must we accept that basic properties of our universe are accidental and uncalculable. In addition, we must believe in the existence of many other universes. But we have no conceivable way of observing these other universes and cannot prove their existence. Thus, to explain what we see in the world and in our mental deductions, we must believe in what we cannot prove.

Sound familiar? Theologians are accustomed to taking some beliefs on faith. Scientists are not. All we can do is hope that the same theories that predict the multiverse also produce many other predictions that we can test here in our own universe. But the other universes themselves will almost certainly remain a conjecture.

"We had a lot more confidence in our intuition before the discovery of dark energy and the multiverse idea," says Guth. "There will still be a lot for us to understand, but we will miss out on the fun of figuring everything out from first principles."

One wonders whether a young Alan Guth, considering a career in science today, would choose theoretical physics.

CHARLES C. MANN

The Birth of Religion

FROM *NATIONAL GEOGRAPHIC*

*The traditional thinking about the origins of human societies is
that the invention of agriculture led to civilization, and civilization
led to religion. Charles C. Mann travels to an 11,000-year-old site
in Turkey, where there is evidence that the conventional view may
get the story backward.*

EVERY NOW AND THEN THE DAWN OF CIVILIZATION IS
reenacted on a remote hilltop in southern Turkey.

The reenactors are busloads of tourists—usually Turkish,
sometimes European. The buses (white, air-conditioned, equipped
with televisions) blunder over the winding, indifferently paved road
to the ridge and dock like dreadnoughts before a stone portal. Visitors flood out, fumbling with water bottles and MP3 players. Guides

call out instructions and explanations. Paying no attention, the visitors straggle up the hill. When they reach the top, their mouths flop open with amazement, making a line of perfect cartoon O's.

Before them are dozens of massive stone pillars arranged into a set of rings, one mashed up against the next. Known as Göbekli Tepe (pronounced Guh-behk-LEE TEH-peh), the site is vaguely reminiscent of Stonehenge, except that Göbekli Tepe was built much earlier and is made not from roughly hewn blocks but from cleanly carved limestone pillars splashed with bas-reliefs of animals—a cavalcade of gazelles, snakes, foxes, scorpions, and ferocious wild boars. The assemblage was built some 11,600 years ago, seven millennia before the Great Pyramid of Giza. It contains the oldest known temple. Indeed, Göbekli Tepe is the oldest known example of monumental architecture—the first structure human beings put together that was bigger and more complicated than a hut. When these pillars were erected, so far as we know, nothing of comparable scale existed in the world.

At the time of Göbekli Tepe's construction much of the human race lived in small nomadic bands that survived by foraging for plants and hunting wild animals. Construction of the site would have required more people coming together in one place than had likely occurred before. Amazingly, the temple's builders were able to cut, shape, and transport 16-ton stones hundreds of feet despite having no wheels or beasts of burden. The pilgrims who came to Göbekli Tepe lived in a world without writing, metal, or pottery; to those approaching the temple from below, its pillars must have loomed overhead like rigid giants, the animals on the stones shivering in the firelight—emissaries from a spiritual world that the human mind may have only begun to envision.

Archaeologists are still excavating Göbekli Tepe and debating its meaning. What they do know is that the site is the most significant in a volley of unexpected findings that have overturned earlier ideas about our species' deep past. Just 20 years ago most researchers believed they knew the time, place, and rough sequence of the Neolithic Revolution—the critical transition that resulted in the birth of agriculture, taking *Homo sapiens* from scattered groups of hunter-gatherers to farming villages and from there to technologically sophisticated societies with great temples and towers and kings and priests who directed the labor of their subjects and recorded their feats in written form. But in recent years multiple new discoveries, Göbekli Tepe preeminent among them, have begun forcing archaeologists to reconsider.

At first the Neolithic Revolution was viewed as a single event—a sudden flash of genius—that occurred in a single location, Mesopotamia, between the Tigris and Euphrates Rivers in what is now southern Iraq, then spread to India, Europe, and beyond. Most archaeologists believed this sudden blossoming of civilization was driven largely by environmental changes: a gradual warming as the Ice Age ended that allowed some people to begin cultivating plants and herding animals in abundance. The new research suggests that the "revolution" was actually carried out by many hands across a huge area and over thousands of years. And it may have been driven not by the environment but by something else entirely.

After a moment of stunned quiet, tourists at the site busily snap pictures with cameras and cell phones. Eleven millennia ago nobody had digital imaging equipment, of course. Yet things have changed less than one might think. Most of the world's great religious centers, past and present, have been destinations for pilgrimages—think of the Vatican, Mecca, Jerusalem, Bodh Gaya (where Buddha was enlightened), or Cahokia (the enormous Native American complex near St. Louis). They are monuments for spiritual travelers, who often came great distances, to gawk at and be stirred by. Göbekli Tepe may be the first of all of them, the beginning of a pattern. What it suggests, at least to the archaeologists working there, is that the human sense of the sacred—and the human love of a good spectacle—may have given rise to civilization itself.

KLAUS SCHMIDT KNEW ALMOST INSTANTLY that he was going to be spending a lot of time at Göbekli Tepe. Now a researcher at the German Archaeological Institute (DAI), Schmidt had spent the autumn of 1994 trundling across southeastern Turkey. He had been working at a site there for a few years and was looking for another place to excavate. The biggest city in the area is Şanlıurfa (pronounced shan-LYOOR-fa). By the standards of a brash newcomer like London, Şanlıurfa is incredibly old—the place where the Prophet Abraham supposedly was born. Schmidt was in the city to find a place that would help him understand the Neolithic, a place that would make Şanlıurfa look young. North of Şanlıurfa the ground ripples into the first foothills of the mountains that run across southern Turkey, source of the famous Tigris and Euphrates Rivers. Nine miles outside of town is a long ridge with a rounded crest that locals call Potbelly Hill—Göbekli Tepe.

In the 1960s archaeologists from the University of Chicago had surveyed the region and concluded that Göbekli Tepe was of little interest. Disturbance was evident at the top of the hill, but they attributed it to the activities of a Byzantine-era military outpost. Here and there were broken pieces of limestone they thought were gravestones. Schmidt had come across the Chicago researchers' brief description of the hilltop and decided to check it out. On the ground he saw flint chips—huge numbers of them. "Within minutes of getting there," Schmidt says, he realized that he was looking at a place where scores or even hundreds of people had worked in millennia past. The limestone slabs were not Byzantine graves but something much older. In collaboration with the DAI and the Şanlıurfa Museum, he set to work the next year.

Inches below the surface the team struck an elaborately fashioned stone. Then another, and another—a ring of standing pillars. As the months and years went by, Schmidt's team, a shifting crew of German and Turkish graduate students and 50 or more local villagers, found a second circle of stones, then a third, and then more.

Geomagnetic surveys in 2003 revealed at least 20 rings piled together, higgledy-piggledy, under the earth.

The pillars were big—the tallest are 18 feet in height and weigh 16 tons. Swarming over their surfaces was a menagerie of animal bas-reliefs, each in a different style, some roughly rendered, a few as refined and symbolic as Byzantine art. Other parts of the hill were littered with the greatest store of ancient flint tools Schmidt had ever seen—a Neolithic warehouse of knives, choppers, and projectile points. Even though the stone had to be lugged from neighboring valleys, Schmidt says, "there were more flints in one little area here, a square meter or two, than many archaeologists find in entire sites."

The circles follow a common design. All are made from limestone pillars shaped like giant spikes or capital T's. Bladelike, the pillars are easily five times as wide as they are deep. They stand an arm span or more apart, interconnected by low stone walls. In the middle of each ring are two taller pillars, their thin ends mounted in shallow grooves cut into the floor. I asked German architect and civil engineer Eduard Knoll, who works with Schmidt to preserve the site, how well designed the mounting system was for the central pillars. "Not," he said, shaking his head. "They hadn't yet mastered engineering." Knoll speculated that the pillars may have been propped up, perhaps by wooden posts.

To Schmidt, the T-shaped pillars are stylized human beings, an idea bolstered by the carved arms that angle from the "shoulders" of some pillars, hands reaching toward their loincloth-draped bellies. The stones face the center of the circle—as at "a meeting or dance," Schmidt says—a representation, perhaps, of a religious ritual. As for the prancing, leaping animals on the figures, he noted that they are mostly deadly creatures: stinging scorpions, charging boars, ferocious lions. The figures represented by the pillars may be guarded by them, or appeasing them, or incorporating them as totems.

Puzzle piled upon puzzle as the excavation continued. For reasons yet unknown, the rings at Göbekli Tepe seem to have regularly lost their power, or at least their charm. Every few decades people buried the pillars and put up new stones—a second, smaller ring, inside the first. Sometimes, later, they installed a third. Then the whole assemblage would be filled in with debris, and an entirely new circle created nearby. The site may have been built, filled in, and built again for centuries.

Bewilderingly, the people at Göbekli Tepe got steadily worse at temple building. The earliest rings are the biggest and most sophisticated, technically and artistically. As time went by, the pillars became smaller, simpler, and were mounted with less and less care. Finally the effort seems to have petered out altogether by 8200 B.C. Göbekli Tepe was all fall and no rise.

As important as what the researchers found was what they did not find: any sign of habitation. Hundreds of people must have been required to carve and erect the pillars, but the site had no water source—the nearest stream was about three miles away. Those workers would have needed homes, but excavations have uncovered no sign of walls, hearths, or houses—no other buildings that Schmidt has interpreted as domestic. They would have had to be fed, but there is also no trace of agriculture. For that matter, Schmidt has found no mess kitchens or cooking fires. It was purely a ceremonial center. If anyone ever lived at this site, they were less its residents than its staff. To judge by the thousands of gazelle and aurochs bones found at the site, the workers seem to have been fed by constant shipments of game, brought from faraway hunts. All of this complex endeavor must have had organizers and overseers, but there is as yet no good evidence of a social hierarchy—no living area reserved for richer people, no tombs filled with elite goods, no sign of some people having better diets than others.

"These people were foragers," Schmidt says, people who gathered plants and hunted wild animals. "Our picture of foragers was always just small, mobile groups, a few dozen people. They cannot make big permanent structures, we thought, because they must move around to follow the resources. They can't maintain a separate class of priests and craft workers, because they can't carry around all the extra supplies to feed them. Then here is Göbekli Tepe, and they obviously did that."

Discovering that hunter-gatherers had constructed Göbekli Tepe was like finding that someone had built a 747 in a basement with an X-Acto knife. "I, my colleagues, we all thought, What? How?" Schmidt said. Paradoxically, Göbekli Tepe appeared to be both a harbinger of the civilized world that was to come and the last, greatest emblem of a nomadic past that was already disappearing. The accomplishment was astonishing, but it was hard to understand how it had been done or what it meant. "In 10 or 15 years," Schmidt predicts, "Göbekli Tepe will be more famous than Stonehenge. And for good reason."

HOVERING OVER GÖBEKLI TEPE IS the ghost of V. Gordon Childe. An Australian transplant to Britain, Childe was a flamboyant man, a passionate Marxist who wore plus fours and bow ties and larded his public addresses with noodle-headed paeans to Stalinism. He was also one of the most influential archaeologists of the past century. A great synthesist, Childe wove together his colleagues' disconnected facts into overarching intellectual schemes. The most famous of these arose in the 1920s, when he invented the concept of the Neolithic Revolution.

In today's terms, Childe's views could be summed up like this: *Homo sapiens* burst onto the scene about 200,000 years ago. For most of the millennia that followed, the species changed remarkably little, with humans living as small bands of wandering foragers. Then came the Neolithic Revolution—"a radical change," Childe said, "fraught with revolutionary consequences for the whole species." In a lightning bolt of inspiration, one part of humankind turned its back on foraging and embraced agriculture. The adoption of farming, Childe argued, brought with it further transformations. To tend their fields, people had to stop wandering and move into permanent villages, where they developed new tools and created pottery. The Neolithic Revolution, in his view, was an explosively important event—"the greatest in human history after the mastery of fire."

Of all the aspects of the revolution, agriculture was the most important. For thousands of years men and women with stone implements had wandered the landscape, cutting off heads of wild grain and taking them home. Even though these people may have tended and protected their grain patches, the plants they watched over were still wild. Wild wheat and barley, unlike their domesticated versions, shatter when they are ripe—the kernels easily break off the plant and fall to the ground, making them next to impossible to harvest when fully ripe. Genetically speaking, true grain agriculture began only when people planted large new areas with mutated plants that did not shatter at maturity, creating fields of domesticated wheat and barley that, so to speak, waited for farmers to harvest them.

Rather than having to comb through the landscape for food, people could now grow as much as they needed and where they needed it, so they could live together in larger groups. Population soared. "It was only after the revolution—but immediately thereafter—that our species really began to multiply at all fast," Childe wrote. In these suddenly more populous societies, ideas could be more readily exchanged, and rates of technological and social innovation soared. Religion and art—the hallmarks of civilization—flourished.

Childe, like most researchers today, believed that the revolution first occurred in the Fertile Crescent, the arc of land that curves northeast from Gaza into southern Turkey and then sweeps southeast into Iraq. Bounded on the south by the harsh Syrian Desert and on the north by the mountains of Turkey, the crescent is a band of temperate climate between inhospitable extremes. Its eastern terminus is the confluence of the Tigris and Euphrates Rivers in southern Iraq—the site of a realm known as Sumer, which dates back to about

4000 B.C. In Childe's day most researchers agreed that Sumer represented the beginning of civilization. Archaeologist Samuel Noah Kramer summed up that view in the 1950s in his book *History Begins at Sumer*. Yet even before Kramer finished writing, the picture was being revised at the opposite, western end of the Fertile Crescent. In the Levant—the area that today encompasses Israel, the Palestinian territories, Lebanon, Jordan, and western Syria—archaeologists had discovered settlements dating as far back as 13,000 B.C. Known as Natufian villages (the name comes from the first of these sites to be found), they sprang up across the Levant as the Ice Age was drawing to a close, ushering in a time when the region's climate became relatively warm and wet.

The discovery of the Natufians was the first rock through the window of Childe's Neolithic Revolution. Childe had thought agriculture the necessary spark that led to villages and ignited civilization. Yet although the Natufians lived in permanent settlements of up to several hundred people, they were foragers, not farmers, hunting gazelles and gathering wild rye, barley, and wheat. "It was a big sign that our ideas needed to be revised," says Harvard University archaeologist Ofer Bar-Yosef.

Natufian villages ran into hard times around 10,800 B.C., when regional temperatures abruptly fell some 12°F, part of a mini ice age that lasted 1,200 years and created much drier conditions across the Fertile Crescent. With animal habitat and grain patches shrinking, a number of villages suddenly became too populous for the local food supply. Many people once again became wandering foragers, searching the landscape for remaining food sources.

Some settlements tried to adjust to the more arid conditions. The village of Abu Hureyra, in what is now northern Syria, seemingly tried to cultivate local stands of rye, perhaps replanting them. After examining rye grains from the site, Gordon Hillman of University College London and Andrew Moore of the Rochester Institute of Technology argued in 2000 that some were bigger than their wild equivalents—a possible sign of domestication, because cultivation inevitably increases qualities, such as fruit and seed size, that people find valuable. Bar-Yosef and some other researchers came to believe that nearby sites like Mureybet and Tell Qaramel also had had agriculture.

If these archaeologists were correct, these protovillages provided a new explanation of how complex society began. Childe thought that agriculture came first, that it was the innovation that allowed humans to seize the opportunity of a rich new environment to extend their dominion over the natural world. The Natufian sites in the Levant suggested instead that settlement came first and that farming arose later, as a product of crisis. Confronted with a drying, cooling environment and growing populations, humans in the remaining fecund areas thought, as Bar-Yosef puts it, "If we move, these other folks will exploit our resources. The best way for us to survive is to settle down and exploit our own area." Agriculture followed.

The idea that the Neolithic Revolution was driven by climate change resonated during the 1990s, a time when people were increasingly worried about the effects of modern global warming. It was promoted in countless articles and books and ultimately enshrined in Wikipedia. Yet critics charged that the evidence was weak, not least because Abu Hureyra, Mureybet, and many other sites in northern Syria had been flooded by dams before they could be fully excavated. "You had an entire theory on the origins of human culture essentially based on a half a dozen unusually plump seeds," ancient-grain specialist George Willcox of the National Center for Scientific Research, in France, says. "Isn't it more likely that these grains were puffed during charring or that somebody at Abu Hureyra found some unusual-looking wild rye?"

As the dispute over the Natufians sharpened, Schmidt was carefully working at Göbekli Tepe. And what he was finding would, once again, force many researchers to reassess their ideas.

ANTHROPOLOGISTS HAVE ASSUMED THAT ORGANIZED religion began as a way of salving the tensions that inevitably arose when hunter-gatherers settled down, became farmers, and developed large societies. Compared to a nomadic band, the society of a village had longer term, more complex aims—storing grain and maintaining permanent homes. Villages would be more likely to accomplish those aims if their members were committed to the collective enterprise. Though primitive religious practices—burying the dead, creating cave art and figurines—had emerged tens of thousands of years earlier, organized religion arose, in this view, only when a common vision of a celestial order was needed to bind together these big, new, fragile groups of humankind. It could also have helped justify the social hierarchy that emerged in a more complex society: Those who rose to power were seen as having a special connection with the gods. Communities of the faithful, united in a common view of the world and their place in it, were more cohesive than ordinary clumps of quarreling people.

Göbekli Tepe, to Schmidt's way of thinking, suggests a reversal of that scenario: The construction of a massive temple by a group of foragers is evidence that organized religion could have come *before* the rise of agriculture and other aspects of civilization. It suggests that the human impulse to gather for sacred rituals arose as humans shifted from seeing themselves as part of the natural world to seeking mastery over it. When foragers began settling down in villages, they unavoidably created a divide between the human realm—a fixed huddle of homes with hundreds of inhabitants—and the dangerous land beyond the campfire, populated by lethal beasts.

French archaeologist Jacques Cauvin believed this change in consciousness was a "revolution of symbols," a conceptual shift that allowed humans to imagine gods—supernatural beings resembling humans—that existed in a universe beyond the physical world. Schmidt sees Göbekli Tepe as evidence for Cauvin's theory. "The animals were guardians to the spirit world," he says. "The reliefs on the T-shaped pillars illustrate that other world."

Schmidt speculates that foragers living within a hundred-mile radius of Göbekli Tepe created the temple as a holy place to gather and meet, perhaps bringing gifts and tributes to its priests and craftspeople. Some kind of social organization would have been necessary not only to build it but also to deal with the crowds it attracted. One imagines chanting and drumming, the animals on the great pillars seeming to move in flickering torchlight. Surely there were feasts; Schmidt has uncovered stone basins that could have been used for beer. The temple was a spiritual locus, but it may also have been the Neolithic version of Disneyland.

Over time, Schmidt believes, the need to acquire sufficient food for those who worked and gathered for ceremonies at Göbekli Tepe may have led to the intensive cultivation of wild cereals and the creation of some of the first domestic strains. Indeed, scientists now believe that one center of agriculture arose in southern Turkey—well within trekking distance of Göbekli Tepe—at exactly the time the temple was at its height. Today the closest known wild ancestors of modern einkorn wheat are found on the slopes of Karaca Dağ, a mountain just 60 miles northeast of Göbekli Tepe. In other words, the turn to agriculture celebrated by V. Gordon Childe may have been the result of a need that runs deep in the human psyche, a hunger that still moves people today to travel the globe in search of awe-inspiring sights.

Some of the first evidence for plant domestication comes from Nevali Çori (pronounced nuh-vah-LUH CHO-ree), a settlement in the mountains scarcely 20 miles away. Like Göbekli Tepe, Nevali Çori came into existence right after the mini ice age, a time archaeologists describe with the unlovely term Pre-pottery Neolithic (PPN). Nevali Çori is now inundated by a recently created lake that provides electricity and irrigation water for the region. But before the waters shut down research, archaeologists found T-shaped pillars and animal images much like those Schmidt would later uncover at Göbekli Tepe. Similar pillars and images occurred in PPN settlements up to a hundred miles from Göbekli Tepe. Much as one can surmise today that homes with images of the Virgin Mary belong to Christians, Schmidt says, the imagery in these PPN sites indicates a shared religion—a community of faith that surrounded Göbekli Tepe and may have been the world's first truly large religious grouping.

Naturally, some of Schmidt's colleagues disagree with his ideas. The lack of evidence of houses, for instance, doesn't prove that nobody lived at Göbekli Tepe. And increasingly, archaeologists studying the origins of civilization in the Fertile Crescent are suspicious of any attempt to find a one-size-fits-all scenario, to single out one primary trigger. It is more as if the occupants of various archaeological sites were all playing with the building blocks of civilization, looking for combinations that worked. In one place agriculture may have been the foundation; in another, art and religion; and over there, population pressures or social organization and hierarchy. Eventually they all ended up in the same place. Perhaps there is no single path to civilization; instead it was arrived at by different means in different places.

In Schmidt's view, many of his colleagues have been as slow to appreciate Göbekli Tepe as he has been to excavate it. This summer will mark his 17th year at the site. The annals of archaeology are replete with scientists who in their hurry carelessly wrecked important finds, losing knowledge for all time. Schmidt is determined not to add his name to the list. Today less than a tenth of the 22-acre site is open to the sky.

Schmidt emphasizes that further research on Göbekli Tepe may change his current understanding of the site's importance. Even its age is not clear—Schmidt is not certain he has reached the bottom layer. "We come up with two new mysteries for every one that we solve," he says. Still, he has already drawn some conclusions. "Twenty years ago everyone believed civilization was driven by ecological forces," Schmidt says. "I think what we are learning is that civilization is a product of the human mind."

JACKSON LEARS

Same Old New Atheism:
On Sam Harris

FROM THE *NATION*

The New Atheism, as it is known, claims to be the only rational alternative to worldviews increasingly dominated by fundamentalist ideologies. In this probing analysis of the work of Sam Harris, one of the best-known New Atheists, Jackson Lears pulls up this "new" philosophy by its poisoned roots.

DURING THE PRESIDENTIAL CAMPAIGN OF 1964, A bit of doggerel surfaced among liberal wits, as they pondered the popularity of Barry Goldwater on certain college campuses:

We're the bright young men,
Who wanna go back to 1910,
We're Barry's boys!
We're the kids with a cause,
A government like granmama's,
We're Barry's boys!

What could be more ludicrous than the spectacle of young people embracing an old reactionary who wanted to repeal the New Deal? One might as well try to revive corsets and spats. Progress in politics, as in other matters, was unstoppable.

These days the satire rings hollow; so too its hubris. Except for the spats, we really have gone back to 1910, if not earlier. The deregulation of business and the starvation of the public sector have returned us to a landscape where irresponsible capital can again roam freely, purchasing legislatures wholesale and trampling on the public interest at will. The Supreme Court has revived the late-nineteenth-century notion that corporations are people, with all the rights of citizenship that personhood entails (including the ability to convert money into free speech). This is a predictable consequence of Republican power, but what is less predictable, and more puzzling, is that the resurrection of Gilded Age politics has been accompanied throughout the culture by a resurgence of Gilded Age patterns of thought, no more so than with the revival of positivism in popular scientific writing.

More a habit of mind than a rigorous philosophy, positivism depends on the reductionist belief that the entire universe, including all human conduct, can be explained with reference to precisely measurable, deterministic physical processes. (This strain of positivism is not to be confused with that of the French sociologist Auguste Comte.) The decades between the Civil War and World War I were positivism's golden age. Positivists boasted that science was on the brink of producing a total explanation of the nature of things, which would consign all other explanations to the dustbin of mythology. Scientific research was like an Easter egg hunt: once the eggs were gathered the game would be over, the complexities of the cosmos reduced to natural law. Science was the only repository of truth, a sovereign entity floating above the vicissitudes of history and power. Science was science.

Though they often softened their claims with Christian rhetoric, positivists assumed that science was also the only sure guide to morality, and the only firm basis for civilization. As their critics began to realize, positivists had abandoned the provisionality of science's experimental outlook by transforming science from a method into a metaphysic, a source of absolute certainty. Positivist assumptions provided the epistemological foundations for Social Darwinism and pop-evolutionary notions of progress, as well as for scientific racism and imperialism. These tendencies coalesced in eugenics, the doctrine that human well-being could be improved and eventually perfected through the selective breeding of the "fit" and the sterilization or elimination of the "unfit."

Every schoolkid knows about what happened next: the catastrophic twentieth century. Two world wars, the systematic slaughter of innocents on an unprecedented scale, the proliferation of unimaginably destructive weapons, brushfire wars on the periphery of empire—all these events involved, in various degrees, the application of scientific research to advanced technology. All showed that science could not be elevated above the agendas of the nation-state: the best scientists were as corruptible by money, power or ideology as anyone else, and their research could as easily be bent toward mass murder as toward the progress of humankind. Science was not merely science. The crowning irony was that eugenics, far from "perfecting the race," as some American progressives had hoped early in the twentieth century, was used by the Nazis to eliminate those they deemed undesirable. Eugenics had become another tool in the hands of unrestrained state power. As Theodor Adorno and Max Horkheimer argued near the end of World War II in *Dialectic of Enlightenment*, the rise of scientific racism betrayed the demonic undercurrents of the positivist faith in progress. Zygmunt Bauman refined the argument forty-two years later in *Modernity and the Holocaust*: the detached positivist worldview could be pressed into the service of mass extermination. The dream of reason bred real monsters.

The midcentury demise of positivism was a consequence of intellectual advances as well as geopolitical disasters. The work of Franz Boas, Claude Lévi-Strauss and other anthropologists promoted a relativistic understanding of culture, which undercut scientific racism and challenged imperial arrogance toward peoples who lagged behind in the Western march of progress. Meanwhile, scientists in disciplines ranging from depth psychology to quantum physics were discovering a physical reality that defied precise definition as well as efforts to reduce it to predictable laws. Sociologists of knowledge, along with historians and philosophers of science (including Karl Mannheim, Peter Berger and Thomas Kuhn), all emphasized the provisionality of scientific truth, its dependence on a shifting expert consensus that could change or even dissolve outright in light of new evidence. Reality—or at least our apprehension of it—could be said to be socially constructed. This meant that our understanding of the physical world is contingent on the very things—the methods of measurement, the interests of the observer—required to apprehend it.

None of this ferment discredited the role of science as a practical means of promoting human well-being: midcentury laboratories produced vaccines and sulfa drugs as well as nuclear weapons. Nor did it prove the existence (or even the possibility) of God, as apologists for religion sometimes claimed. But it did undermine the positivist faith in science as a source of absolute certainty and moral good. As ethical guides, scientists had proved to be no more reliable than anyone else. Apart from a few Strangelovian thinkers (the physicist Edward Teller comes to mind), scientists retreated from making ethical or political pronouncements in the name of science.

DURING THE PAST SEVERAL DECADES, there has been a revival of positivism alongside the resurgence of laissez-faire economics and other remnants of late-nineteenth-century social thought. E. O. Wilson's *Sociobiology* (1975) launched pop-evolutionary biologism on the way to producing "evolutionary psychology"—a parascience that reduces complex human social interactions to adaptive behaviors inherited from our Pleistocene ancestors. Absence of evidence from the Pleistocene did not deter evolutionary psychologists from telling Darwinian stories about the origins of contemporary social life. Advances in neuroscience and genetics bred a resurgent faith in the existence of something called human nature and the sense that science is on the verge of explaining its workings, usually with reference to brains that are "hard-wired" for particular kinds of adaptive, self-interested behavior. In the problematic science of intelligence testing, scientific racism made a comeback with the publication of Richard Herrnstein and Charles Murray's *The Bell Curve* in 1994.

This resurgent positivism provoked ferocious criticism, most of it serious and justified. Stephen Jay Gould took dead aim at what he called "Darwinian Fundamentalism," arguing that strict adaptationist accounts of evolutionary thought presented "a miserly and blinkered picture of evolution," impoverished not only by the lack of evidence but also by the reductionist tendency to insist on the simplest possible explanation for the complexities of human and animal behavior. Other critics—Noam Chomsky, Richard Lewontin—joined Gould in noting the tendency of Darwinian fundamentalists to "prove" adaptationist arguments by telling "just-so stories." These are narratives about evolution based on hypotheses that are plausible, and internally consistent with the strict adaptationist program, but lacking the essential component of the scientific method: falsifiability. This was a powerful argument.

Within the wider culture, however, reductionism reigned. Hardly a day went by without journalists producing another just-so story about primitive life on the savanna thousands of years ago, purporting to show why things as they are have to be the way they are. In these stories, the parched fruits of a mirthless and minor imagination, all sorts of behavior, from generals' exaggerations of their armies' strength to the promiscuity of powerful men, could be viewed as an adaptive strategy, embedded in a gene that would be passed on to subsequent generations. In the late twentieth century, as in the late nineteenth, positivism's account of human behavior centered on the idea that the relentless assertion of advantage by the strong serves the evolutionary interests of the species. Positivism remained a mighty weapon of the status quo, ratifying existing arrangements of wealth, power and prestige.

The terrorist attacks of September 11, 2001, injected positivism with a missionary zeal. "Once I had experienced all the usual mammalian gamut of emotions, from rage to nausea, I also discovered that another sensation was contending for mastery," Christopher Hitchens wrote several months after 9/11. "On examination, and to my own surprise and pleasure, it turned out to be exhilaration. Here was the most frightful enemy—theocratic barbarism—in plain view. . . . I realized that if the battle went on until the last day of my life, I would never get bored in prosecuting it to the utmost." Putting aside the question of how Hitchens intended to "prosecute" this battle other than pontificating about it, and the irrelevance of his boredom to dead and maimed soldiers and civilians, one cannot deny that he embraced, from a safe distance, the "war on terror" as an Enlightenment crusade. He was not alone. Other intellectuals fell into line, many holding aloft the banner of science and reason against the forces of "theocratic barbarism." Most prominent were the intellectuals the media chose to anoint, with characteristic originality, as the New Atheists, a group that included Hitchens, Daniel Dennett, Richard Dawkins and Sam Harris. In the shadow of 9/11, they were ready to press the case against religion with renewed determination and fire.

Atheism has always been a tough sell in the United States. In Europe, where for centuries religious authority was intertwined with government power, atheists were heroic dissenters against the unholy alliance of church and state. In the United States, where the two realms are constitutionally separate, Protestant Christianity suffused public discourse so completely in the late nineteenth and early twentieth centuries that some positivists felt the need to paper over their differences with religion. US politics has frequently been flooded by waves of Christian fervor. Sometimes religion has bolstered the forces of political sanctimony and persecution, as with Prohibition in the 1920s and anticommunism during the cold war; but it has also encouraged dissenters to speak truth to power—to abolish slavery, to regulate capitalism, to end the Vietnam War.

The Christian right, which had risen to prominence in the late twentieth century, provided an unprecedented target for New Atheists' barbs. Here was a particularly noxious form of religion in American politics—more dangerous than the bland piety of politicians or the ad nauseam repetition of "God Bless America." From the Reagan administration to that of George W. Bush, the Christian right succeeded in shifting political debates from issues of justice and equality to moral and cultural questions, often persuading working-class voters to cast ballots for candidates whose policies undercut their economic interests. Rage about abortion and same-sex marriage drowned out discussion of job security and tax equity. Fundamentalist Christians denied global warming and helped to derail federal funding for stem-cell research. Most catastrophically, they supplied the language of Providence that sanctified Bush's "war on terror" as a moral crusade.

Still, it remains an open question how much this ideological offensive depended on religious dogma, and how much it was the work of seasoned political players, such as plutocrats bent on deregulating business and dismantling progressive taxation, corporate-sponsored media eager to curry favor with the powerful and military contractors hoping to sup at the public trough. Even the rhetoric of Providential mission owed more to romantic nationalism than to orthodox Christianity, which has long challenged the cult of the nation-state as a form of idolatry.

THE NEW ATHEISTS DID NOT bother with such nuance. Hitchens and Harris, in particular, wasted no time enlisting in Bush's crusade, which made their critique of religion selective. It may have targeted Christianity and occasionally Judaism, but hatred and fear of Islam was its animating force. Despite their disdain for public piety, the New Atheists provided little in their critique to disturb the architects and proselytizers of American empire: indeed, Hitchens and Harris asserted a fervent rationale for it. Since 9/11, both men have made careers of posing as heroic outsiders while serving the interests of the powerful.

Of the two, Harris has the more impressive credentials. In addition to being a prolific pundit on websites, a marquee name on the lecture circuit and the author of three popular books, *The End of Faith* (2004), *Letter to a Christian Nation* (2006) and *The Moral Landscape* (2010), he is a practicing neuroscientist who emerges from the lab to reveal the fundamental truths he claims to have learned there. Chief among them are the destructive power of religion, which Harris always defines in the most literal and extreme terms, and the immediate global threat of radical Islam. Everything can be explained by the menace of mobilized religious dogma, which is exacerbated by liberal tolerance. Stupefied by cultural relativism, we refuse to recognize that some ways of being in the world—our own especially—are superior to others. As a consequence, we are at the mercy of fanatics who will stop at nothing until they "refashion the societies of Europe into a new Caliphate." They are natural-born killers, and we are decadent couch potatoes. Our only defense, Harris insists, is the rejection of both religion and cultural relativism, and the embrace of science as the true source of moral value.

Harris claims he is committed to the reasonable weighing of evidence against the demands of blind faith. This is an admirable stance, but it conceals an absolutist cast of mind. He tells us that because "the well-being of conscious [and implicitly human] creatures" is the only reliable indicator of moral good, and science the only reliable means for enhancing well-being, only science can be a source of moral value. Experiments in neuroimaging, Harris argues, reveal that the brain makes no distinction between judgments of value and judgments of fact; from this finding he extracts the non sequitur that fact and value are the same. We may not know all the moral truths that research will unearth, but we will soon know many more of them. Neuroscience, he insists, is on the verge of revealing the keys to human well-being: in brains we trust.

To define science as the source of absolute truth, Harris must first ignore the messy realities of power in the world of Big Science. In his books there is no discussion of the involvement of scientists in the military-industrial complex or in the pharmacological pursuit of profit. Nor is any attention paid to the ways that chance, careerism and intellectual fashion can shape research: how they can skew data, promote the publication of some results and consign others to obscurity, channel financial support or choke it off. Rather than provide a thorough evaluation of evidence, Harris is given to sweeping, unsupported generalizations. His idea of an argument about religious fanaticism is to string together random citations from the Koran or the Bible. His books display a stunning ignorance of history, including the history of science. For a man supposedly committed to the rational defense of science, Harris is remarkably casual about putting a thumb on the scale in his arguments.

If we evaluate those arguments according to their resonance with public policy debates, the results are sobering. Harris's convictions reveal his comfortable cohabitation with imperial power. From him we learn, among other things, that torture is just another form of collateral damage in the "war on terror"—regrettable, maybe, but a necessary price to pay in the crucial effort to save Western civilization from the threat of radical Islam. We also learn that pacifism, despite its (allegedly) high moral standing, is "immoral" because it leaves us vulnerable to "the world's thugs." As in the golden age of positivism, a notion of sovereign science is enlisted in the service of empire. Harris dispenses with the Christian rhetoric of his imperialist predecessors but not with their rationalizations for state-sponsored violence. Posing as a renegade on the cutting edge of scientific research and moral enlightenment, Harris turns out to be one of the bright young men who want to go back to 1910.

The End of Faith, WRITTEN in the wake of 9/11, bears all the marks of that awful time: hysteria, intolerance, paranoia; cankered demands for unity and the demonization of dissent. The argument is simple: the attacks on the World Trade Center awakened us to the mortal danger posed by dogmatic religion. Enlightened atheists must take up Voltaire's challenge and crush the infamous thing at last—with the weight of scientific arguments if possible, with the force of military might if necessary. Though *The End of Faith* includes a chapter of complaint about the Christian right and Bush's God-intoxicated White House, Harris singles out Islam as his enemy: "Anyone who says that the doctrines of Islam have 'nothing to do with terrorism' . . . is just playing a game with words."

The politics of Harris's argument are rooted in the Manichaean moralism of Samuel Huntington's 1993 article in *Foreign Affairs* about the "clash of civilizations" between the West and an emerging "Islamic-Confucian" civilization. Huntington may have been wrong about the Confucian element, but his apocalyptic dualism fed the revenge fantasies of the post-9/11 United States. Harris endorses Huntington's argument uncritically, with characteristic indifference to historical evidence: "One need only read the Koran to know," he tells us, that Huntington was right. I am reminded of my fellow naval officers' insistence, during the Vietnam War, that one need only read *The Communist Manifesto* to ascertain the Kremlin's blueprint for world domination.

Harris's tunnel vision leads him to overlook the roots of radical Islam, including the delusion of a revived caliphate, in the twentieth-century politics of imperial rivalries and anti-imperial resistance. (Indeed, under scrutiny, Islamic jihad is looking less like a revolutionary religious movement and more like the guerrilla fantasy of some angry young Arab men—educated, unemployed and humiliated by actual or imagined imperial arrogance. Radical Islam often provides an idiom for their anger, but its centrality has been exaggerated.) Terrorism is not linked to poverty, oppression or humiliation, Harris insists: the world is full of poor people who are not terrorists. Terrorism is the rough beast of Islam, which is "undeniably a religion of conquest." Our choices are clear: "The West must either win the argument [with Muslim orthodoxy] or win the war. All else will be bondage." Ironically, "the only thing that currently stands between us and the roiling ocean of Muslim unreason is a wall of tyranny and human rights abuses [in Arab countries] that we have helped to erect." It is time to remake the Middle East in the name of science and democracy, to convert the Muslim believers to unbelief and save them from themselves. The recent, extraordinary revolution in Egypt, a nationwide, nonsectarian call for democratic reform and a more equitable distribution of resources, underscores the provincial arrogance of this perspective.

But the intellectual problems run deeper. The conceptual muddle at the core of Harris's argument is directly traceable to Huntington's essay. Groping for a global conflict to replace the recently ended cold war, Huntington fell into the fatal error of confusing civilizations with nations. As William Pfaff reminds us, "Islamic civilization is huge":

Nearly all of the Muslim nations except Iran . . . conduct normal political and economic relations with most if not all of the Western countries. The notion that the members of this global religious civilization are at "war" with Western civilization, or are vulnerable to political radicalization by a few thousand Arab mujahideen because of Middle Eastern and South Asian political issues—of which most of the global Muslim population knows little—is a Western fantasy.

Fantastic as it is, the vision of Armageddon appeals to the longing for clarity and certainty at the heart of the positivist sensibility. "All pretensions to theological knowledge should now be seen from the perspective of a man who was just beginning his day on the one hundredth floor of the World Trade Center on the morning of September 11, 2001," Harris writes. That is a pretty limited perspective.

Harris is as narrow in his views as the believers he condemns. Consider his assault on "the demon of relativism," which, he declares, leaves us unprepared to face our ignorant tribal adversaries and robs us of the moral resources needed to prevail in the Armageddon against unreason. This conviction stems from a profound ignorance of philosophy. Harris finds it "interesting" that Sayyid Qutb, Osama bin Laden's favorite thinker, felt that philosophical pragmatism "would spell the death of American civilization." Pragmatism causes its devotees "to lose the conviction that you can actually be right—about *anything*," Harris announces. One can only imagine the astonishment of pragmatists such as William James, who opposed America's imperial adventures in Cuba and the Philippines, or John Dewey, a staunch defender of progressive education, if told that their inclination to evaluate ideas with respect to their consequences somehow prevented them from holding convictions. For Harris, pragmatism and relativism undermine the capacity "to admit that not all cultures are at the same stage of moral development," and to acknowledge our moral superiority to most of the rest of the world. By preventing us from passing judgment on others' beliefs, no matter how irrational, "religious tolerance" has become "one of the principal forces driving us toward the abyss." Harris treats the recognition of legitimate moral differences as a sign of moral incompetence, and it is this sort of posturing that has cemented the New Atheists' reputation for bold iconoclasm.

HARRIS'S ARGUMENT AGAINST RELATIVISM IS muddled and inconsistent on its own terms, but it is perfectly consistent with the aims of the national security state. It depends on the assumption that Americans (and "the West") exist on a higher moral plane than just about anyone else. "As a culture, we have clearly outgrown our tolerance for the deliberate torture and murder of innocents," Harris writes in *The End of Faith*. "We would do well to realize that much of the world has not." He dismisses equations of state-sponsored violence (which creates collateral damage) and terrorist violence (which deliberately targets civilians): "Any honest witness to current events will realize that there is no moral equivalence between the kind of force civilized democracies project in the world, warts and all, and the internecine violence that is perpetrated by Muslim militants, or indeed by Muslim governments." He asks critics of civilian casualties in the Iraq War to imagine if the situation were reversed, and the Iraqi Republican Guard had invaded Washington. Do they think Iraqis would have taken as great care to spare civilians as the Americans did? "We cannot ignore human intentions. Where ethics are concerned, intentions are everything."

One would think that Harris's intentionalism would have him distinguish between the regrettable accidents of collateral damage and the deliberate cruelty of torture. But after invoking a series of fantastic scenarios ranging from the familiar ticking time bomb to demonic killers preparing to asphyxiate 7-year-old American girls, Harris concludes that the larger intentions animating torture can be as noble as those that cause collateral damage: there is "no ethical difference" between them, he says. Torture, from this bizarrely intentionalist view, is somehow now a form of collateral damage. Both are necessary tactics in a fight to the death against Islamic unreason. "When your enemy has no scruples, your own scruples become another weapon in his hand," Harris writes. "We cannot let our qualms over collateral damage paralyze us because our enemies know no such qualms." Most treacherous are the qualms of pacifists, whose refusal to fight is really "nothing more than a willingness to die, and to let others die, at the pleasure of the world's thugs." (Reading this passage, one can't help wondering why in 2005 PEN bestowed its Martha Albrand Award for First Nonfiction upon *The End of Faith*.) Given the implacable opposition between Islam and Western modernity, "it seems certain that collateral damage, of various sorts, will be a part of our future for many years to come." It is the endless war against evil, the wet dream of every armchair combatant from Dick Cheney to Norman Podhoretz.

The only difference is that, unlike those pious gents, Harris dismisses not only Islam but also all the Western monotheisms as "dangerously retrograde" obstacles to the "global civilization" we must create if we are to survive. His critique of religion is a stew of sophomoric simplifications: he reduces all belief to a fundamentalist interpretation of sacred texts, projecting his literalism and simple-mindedness onto believers whose faith may foster an epistemology far more subtle than his positivist convictions. Belief in scriptural inerrancy is Harris's only criterion for true religious faith. This eliminates a wide range of religious experience, from pain and guilt to the exaltation of communal worship, the ecstasy of mystical union with the cosmos and the ambivalent coexistence of faith and doubt.

But Harris is not interested in religious experience. He displays an astonishing lack of knowledge or even curiosity about the actual content of religious belief or practice, announcing that "most religions have merely canonized a few products of ancient ignorance and derangement and passed them down to us as though they were primordial truths." Unlike medicine, engineering or even politics, religion is "the mere maintenance of dogma, is one area of discourse that does not admit of progress." Religion keeps us anchored in "a dark and barbarous past," and what is generally called sacred "is not sacred for any reason other than that it was thought sacred *yesterday*." Harris espouses the Enlightenment master narrative of progress, celebrating humans' steady ascent from superstition to science; no other sort of knowledge, still less wisdom, will do.

There is one religious practice Harris does admit to tolerating: Buddhist meditation, which allows one to transcend mind-body dualism and view the self as process. Only the wisdom of the East offers any access to this experience of self, Harris insists, as he tosses off phrases plucked at random from a Zen handbook. Given the persistent popularity of the wisdom of the East among the existential homeless of the West, the exemption Harris grants Buddhism is perfectly predictable, as is his thoroughgoing ignorance of Western intellectual tradition. "Thousands of years have passed since any Western philosopher imagined that a person should be made happy, peaceful, or even wise, in the ordinary sense, by his search for truth," Harris proclaims, ignoring Montaigne, Erasmus, Ignatius of Loyola, Thomas Merton, Martin Buber, Meister Eckhart and a host of other Protestants, Catholics, Jews and humanists. Harris's lack of curiosity complements his subservience to cultural fashion.

SIMILAR WEAKNESSES ABOUND IN *Letter to a Christian Nation*, in which Harris taunts the many Christians infuriated by his first book. Harris admits up front that "the 'Christian' I address throughout is a Christian in a narrow sense of the term." Aiming comfortably at this caricature, he repeats his insistence that there is a fatal clash of civilizations afoot, between Islam and the West but also between science and religion. Armageddon still looms.

This screed is striking only because it affirms Harris's positivistic fundamentalism with exceptional clarity. "When considering the truth of a proposition," he writes, "one is either engaged in an honest appraisal of the evidence and logical arguments, or one isn't." But consider the ambiguities of statistical research in Harris's field of brain science, in particular the difficulty psychologists and neuroscientists have had in replicating results over time—a development recently surveyed by Jonah Lehrer in *The New Yorker*. Lehrer discusses the research of the psychologist Jonathan Schooler, who revealed that describing a face makes recognizing it more difficult rather than easier. This phenomenon is called the "verbal overshadowing" effect, which became big news in the scientific study of memory, and made Schooler's career. But over time, and despite scrupulous attention to detail and careful design of his experiments, Schooler found it increasingly difficult to replicate his earlier findings. His research had been run aground by the baffling "decline effect" that scientists have struggled with for decades, a result (or nonresult) that suggests that there may be disturbing limitations to the scientific method, at least in the statistically based behavioral sciences.

Some decline effects can arise from less mysterious sources, beginning with the vagaries of chance and the statistical drift toward the mean. But in other cases, Lehrer explains, statistical samples can change over time. Drugs that have passed clinical trials—such as the "second-generation antipsychotics" Abilify, Seroquel and Zyprexa—can be initially tested on schizophrenics, then prescribed to people with milder symptoms for whom they are less effective. Conceptual foundations for research can also be shaky, such as the notion that female swallows choose male mates on the basis of their "symmetry." The questions arise: How does one precisely measure a symmetrical arrangement of feathers? At what point does symmetry end and asymmetry begin? These sorts of problems make replicating results more difficult, and the difficulties are compounded by the standard practices of professional science. Initial research success is written up for scientific journals, rewarded with grants and promotions, and reported to credulous nonscientists; subsequent failures to replicate results remain largely invisible—except to the researchers, who, if they are honest in their appraisal of the evidence, find it hard to accept simple-minded notions of statistically based certainty. The search for scientific truth is not as straightforward as Harris would like to believe.

Methodological and professional difficulties of this sort do not clutter *The Moral Landscape*, Harris's recent effort to fashion a science of ethics. Incredibly, nearly a decade after 9/11, Harris continues to dwell on the fear of Muslim extremists establishing a new caliphate across Europe, making unreason the law of the land and forcing Parisian shopgirls to wear burqas. In looking for examples of religious barbarism, Harris always turns first to what he calls "the especially low-hanging fruit of conservative Islam." But his main target in this book is the "multiculturalism, moral relativism, political correctness, tolerance even of *intolerance*" that hobbles "the West" in its war against radical Islam.

He is especially offended by anthropology. Too often, he says, "the fire-lit scribblings of one or another dazzled ethnographer" have sanctioned some destructive practice (human sacrifice, female genital mutilation) by explaining its adaptive or social function. At their worst, ethnographers have created a cult of the noble savage that celebrates primitive cultures we should rightfully scorn. His scornfulness aside, Harris is not wrong about ethnographic sentimentality, but he thoroughly misunderstands cultural relativism. He seems to think it means cultivating a bland indifference to ethical questions rather than making a serious effort to understand ethical perspectives radically different from our own without abandoning our own. He is ignorant of the relevant anthropological literature on the subjects that vex him the most, such as Hanna Papanek's study of Pakistani women, which described the burqa as "portable seclusion," a garment that allowed women to go out into the world while protecting them from associating with unrelated men. As the anthropologist Lila Abu-Lughod writes, the burqa is a "mobile home" in patriarchal societies where women are otherwise confined to domestic space. Harris cannot imagine that Islamic women might actually choose to wear one; but some do. Nor is he aware of the pioneering work of Christine Walley on female genital mutilation in Africa. Walley illuminates the complex significance of the practice without ever expressing tolerance for it, and she uses cross-cultural understanding as a means of connecting with local African women seeking to put an end to it.

Harris's version of scientific ethics does not allow for complexity. In *The Moral Landscape*, he describes his philosophical position as a blend of moral realism ("moral claims can really be true or false") and consequentialism ("the rightness of an act depends on how it impacts the well-being of conscious creatures"). He does not explain why he has abandoned the intentionalism he espoused in *The End of Faith*. Nor does he spell out how his newfound consequentialism can allow him to maintain his justification of collateral damage (which surely "impacts the well-being of conscious creatures"), or how his new view differs from the pragmatism he had previously condemned. Pragmatism, the argument that ideas become true or false as their impact on the world unfolds, is nothing if not consequentialist.

COMPETING PHILOSOPHICAL CLAIMS MERGE, FOR Harris, in "the moral brain." Moral truth is not divine in origin, nor is it merely a product of "evolutionary pressure and cultural invention." It is a scientific fact. Or it soon will be: "The world of measurement and the world of meaning must eventually be reconciled." This is not an argument for Western ethnocentrism, Harris insists, but rather for the idea "that the most basic facts about human flourishing must transcend culture, just as most other facts do." No one can dispute the desirability of human flourishing or the possibility that neuroscience may lead us closer to it. But the big questions always lead outward from the brain to the wider world. If altruism has an innate biological basis, as some research suggests, how can societies be made to enhance it rather than undermine it?

Harris's reductionism leads him in the opposite direction. His confidence in scientific ethics stems from his discovery that "beliefs about facts and beliefs about values seem to arise from similar processes at the level of the brain." Much of *The Moral Landscape* is devoted to teasing inferences from this finding. Sometimes Harris merely belabors the obvious. For instance, he points out that the medial prefrontal cortex (MPFC), which records feelings of reward and "self-relevance," also registers the difference between belief and disbelief. When research subjects are presented with a moral dilemma—to save five people by killing one—the prospect of direct personal involvement more strongly activates brain regions associated with emotion. As Harris observes, "pushing a person to his death is guaranteed to traumatize us in a way that throwing a switch will not." We do not need neuroscience to confirm the comparative ease of killing at a distance: Bauman's work on the Holocaust, along with many other studies, demonstrated this decades ago.

More commonly, though, Harris depends on the MPFC to make more provocative claims. He says nothing about the pool of test subjects or the methods used to evaluate evidence in these experiments. Instead he argues by assertion. As he writes, "involvement of the MPFC in belief processing . . . suggests that the physiology of belief may be the same regardless of a proposition's content. It also suggests that the division between facts and values does not make much sense in terms of underlying brain function." This is uncontroversial but beside the point. The nub of the matter is not the evaluation of the fact-value divide "in terms of underlying brain function" but the conscious fashioning of morality. Harris is undaunted. He asks, "If, from the point of view of the brain, believing 'the sun is a star' is importantly similar to believing 'cruelty is wrong,' how can we say that scientific and ethical judgments have nothing in common?" But can the brain be said to have a "point of view"? If so, is it relevant to morality?

There is a fundamental reductionist confusion here: the same biological origin does not constitute the same cultural or moral significance. In fact, one could argue, Harris shows that the brain cannot distinguish between facts and values, and that the elusive process of moral reasoning is not reducible to the results of neuroimaging. All we are seeing, here and elsewhere, is that "brain activity" increases or decreases in certain regions of the brain during certain kinds of experiences—a finding so vague as to be meaningless. Yet Harris presses forward to a grandiose and unwarranted conclusion: if the fact-value distinction "does not exist as a matter of human cognition"—that is, measurable brain activity—then science can one day answer the "most pressing questions of human existence": Why do we suffer? How can we be happy? And is it possible to love our neighbor as ourselves?

These high-minded questions conceal a frightening Olympian agenda. Harris is really a social engineer, with a thirst for power that sits uneasily alongside his allegedly disinterested pursuit of moral truth. We must use science, he says, to figure out why people do silly and harmful things in the name of morality, what kinds of things they should do instead and how to make them abandon their silly and harmful practices in order "to live better lives." Harris's engineering mission envelops human life as a whole. "Given recent developments in biology, we are now poised to consciously engineer our further evolution," he writes. "Should we do this, and if so, in which ways? Only a scientific understanding of the possibilities of human well-being could guide us." Harris counsels that those wary of the arrogance, and the potential dangers, of the desire to perfect the biological evolution of the species should observe the behavior of scientists at their professional meetings: "arrogance is about as common at a scientific conference as nudity." Scientists, in Harris's telling, are the saints of circumspection.

IF THAT'S TRUE, THEN HARRIS breaks the mold. Nowhere is this clearer, or more chilling, than in his one extended example of a specific social change that could be effected by scientific ethics. Convinced that brain science has located the biological sources of "bias"—the areas of the brain that cause us to deviate from the norms of factual and moral reasoning—Harris predicts that this research will lead to the creation of foolproof lie detectors. He does not say how these devices will be deployed. Will they be worn on the body, implanted in the brain, concealed in public locations? What he does say is that they will be a great leap forward to a world without deception—which, we must understand, is one of the chief sources of evil. "Whether or not we ever crack the neural code, enabling us to download a person's private thoughts, memories, and perceptions without distortion," he declares, the detectors will "surely be able to determine, to a moral certainty, whether a person is representing his thoughts, memories, and perceptions honestly in conversation." (As always, the question arises, who are "we"?) Technology will create a brave new world of perfect transparency, and legal scholars who might worry about the Fifth Amendment implications are being old-fashioned. The "prohibition against compelled testimony itself appears to be a relic of a more superstitious age," Harris writes, when people were afraid "that lying under oath would damn a person's soul for eternity." He does admit that because "no technology is ever perfect," it's likely that a few innocent people will be condemned; but the courts do that already, he notes, and besides, deception will have become obsolete. Rarely in all his *oeuvre* has Harris's indifference to power and its potential abuse been more apparent or more abominable.

Maybe this explains why Harris remains an optimist despite all the "dangerously retrograde" orthodoxies on the loose. Moral progress is unmistakable, he believes, at least in "the developed world." His chief example is how far "we" have moved beyond racism. Even if one accepts this flimsy assertion, the inconvenient historical fact is that, intellectually at least, racism was undone not by positivistic science, which underwrote it, but by the cultural relativism Harris despises. Ultimately his claims for moral progress range more widely, as he reports that "we" in "the developed world" are increasingly "disturbed by our capacity to do one another harm." What planet does this man live on? Besides our wars in Afghanistan and Iraq, "we" in the United States are engaged in a massive retreat from the welfare state and from any notion that we have a responsibility to one another or to a larger public good that transcends private gain. This retreat has little to do with Islamic radicalism or the militant piety of the Christian right, though the latter does remain a major obstacle to informed debate. The problem in this case is not religion. Despite the fundamental (or perhaps even innate) decency of most people, our political and popular culture does little to encourage altruism. The dominant religion of our time is the worship of money, and the dominant ethic is "To hell with you and hooray for me."

Harris is oblivious to this moral crisis. His self-confidence is surpassed only by his ignorance, and his writings are the best argument against a scientific morality—or at least one based on his positivist version of science and ex cathedra pronouncements on politics, ethics and the future of humanity. In *The Moral Landscape* he observes that people (presumably including scientists) often acquire beliefs about the world for emotional and social rather than cognitive reasons: "It is also true that the less competent a person is in a given domain, the more he will tend to overestimate his abilities. This often produces an ugly marriage of confidence and ignorance that is very difficult to correct for." The description fits Harris all too aptly, as he wanders from neuroscience into ethics and politics. He may well be a fine neuroscientist. He might consider spending more time in his lab.

RACHEL AVIV

God Knows Where I Am

FROM THE *NEW YORKER*

Through the story of a mentally ill woman who chronicled her last days in harrowing detail, Rachel Aviv uncovers the gaps and inadequacies in the American mental-health-care system that too often lead not to treatment but tragedy.

O N OCTOBER 5, 2007, TWO DAYS AFTER BEING RE-leased from New Hampshire Hospital, in Concord, Linda Bishop discarded all her belongings except for mascara, tweezers, and a pen. For nearly a year, she had complained about the restrictions of the psychiatric unit, but her only plan for her release was to remain invisible. She spent two nights in a field she called Hoboville, where homeless people slept, and then began wandering around Concord, avoiding the main streets. Wary of spies, she cut through the underbrush behind buildings, walked through gullies beside the roads, and, when she needed to rest, huddled in the bushes. Her life was saved along the way, she later wrote, by two warblers and an owl.

A tall, athletic fifty-one-year-old with blue eyes and a bachelor's degree in art history from the University of New Hampshire, Linda had been admitted to the hospital in late October, 2006, after having been found incompetent to stand trial for a series of offenses. She spent most of her eleven months there reading, writing, and crocheting. She refused all psychiatric medication, because she believed her diagnosis (bipolar disorder with psychosis) was a mistake. Each time she met a new psychiatrist, she declared her lack of respect for the profession. Only when conversations moved away from her mental illness, a term she generally placed in quotation marks, was she cheerful and engaged. Her medical records consistently note the same traits: "extremely bright," "very pleasant," "denies completely that she has an illness." In the weeks leading up to her discharge, her doctors urged her to make arrangements for housing and follow-up care, but Linda refused, saying, "God will provide."

During a rainstorm on her fourth day out of the hospital, Linda broke into a vacant farmhouse for sale on Mountain Road, a scenic residential street. The three-story home overlooked a brook and an apple orchard, and a few rooms were still sparsely furnished. Linda intended to stay only a few nights, but she began to worry that her dirty clothes would attract attention if she walked back to town. "I look terrible . . . like a vagrant," she wrote in a black leather pocket notebook that the previous tenants had left behind. Linda had led a nomadic existence ever since she had abandoned her sleeping thirteen-year-old daughter, in 1999, leaving a note saying that she was going to meet the governor. She drifted between homeless shelters, hospitals, and jail. She wrote in the journal that she wasn't ready to "make my presence known—and just start the whole mess again—to prove what—that I'm all right? Have done that too many times." Two days after breaking into the house, she decided to make the place her temporary home. She would subsist on apples while "awaiting further instructions" from God.

Linda settled into a routine. In the morning, when the sun poured through the living-room window, warming the end of the couch, she read college textbooks she found in the attic. The former tenant appeared to have dropped out of school in 1969 ("but his creative writing is very good!" she noted), and she began embarking on the education he had abandoned. She began with Joseph Conrad and moved on to biology ("chloroplasts, lysosomes, mitochondria + cell division!") and *Great Issues in Western Civilization*. When she had enough energy, she did her "chores." She combed her graying brown hair—first with a small rake, and, when that proved too cumbersome, with a fork—and tidied the house, in case potential buyers came for a viewing. There was no electricity or water, but, after dusk, she rinsed her underwear in the brook, collected water with a vase, and picked apples.

After the first week, she estimated that she had lost ten pounds. When she looked in the mirror, she was startled by how drawn her face had become. Yet after enduring so many irritations in her hospital unit—patients who wouldn't stop talking, or who touched her, or sat in her favorite chair, or made noise in the middle of the night—she didn't mind having time alone. From her windows, she enjoyed watching purple finches, tufted titmice, chickadees, and "Mr. and Mrs. Cardinal." She wished she had binoculars. A neighbor came over to mow the lawn and pull the weeds. "He has no idea I'm here!" Linda wrote, as she watched him from an upstairs window.

The threat that Linda was hiding from was a shifty one—she alluded to conspiracies involving her older sister, the government, and Satan's workers—but she also wondered if anyone was even looking for her. She kept retracing the series of events that had led her to this house. She knew it didn't "make sense to be barely existing"—she got light-headed just walking up the stairs—but she felt that the situation must have been willed by the Lord. By the end of October, she had a stash of three hundred apples. She worried about the coming winter as she watched trees lose their leaves, milkweed seeds blow in the wind "like it's snowing," and geese migrate south. Still, she could find "no signs or clues that I should be doing anything different."

THROUGHOUT LINDA'S STAY AT NEW Hampshire Hospital, her doctors routinely noted that she lacked "insight," a term that has a troubled legacy in psychiatry. Studies have shown that nearly half of people given a diagnosis of psychotic illness, such as schizophrenia or bipolar disorder, say that they are not mentally ill—naturally, they also tend to resist treatment. The psychiatrist Aubrey Lewis defined insight in 1934 in the *British Journal of Medical Psychology* as the "correct attitude to a morbid change in oneself." But the definition was so ambiguous that his paper was ignored for over fifty years. Psychiatrists were reluctant to move away from objective, observable phenomena and to examine the private ways that people make sense of the experience of losing their minds. Today, insight is assessed every time a patient enters a psychiatric hospital, through the Mental Status Examination, but this form of awareness is still poorly understood. Patients are considered insightful when they can reinterpret unusual occurrences—growing angel's wings, feeling as if their organs have been removed, decoding political messages in street signs—as psychiatric symptoms. In the absence of any clear neurological marker of psychosis, the field revolves around a paradox: an early sign of sanity is the ability to recognize that you've been insane. (A "correct attitude," for most Western psychiatrists, would exclude interpretations featuring spirits, demons, or karmic disharmony.)

Getting patients to acknowledge their own disorders also has become an ethical imperative. Implicit in the doctrine of informed consent is the notion that before agreeing to take medication patients should be aware of the nature and course of their own illnesses. In balancing rights against needs, though, psychiatry is stuck in a kind of moral impasse. It is the only field in which refusal of treatment is commonly viewed as a manifestation of illness rather than as an authentic wish. According to Linda's treatment review, her most perplexing behavior was her "continuing denial of the legitimacy of her 'patienthood.'"

When psychoanalytic theories were dominant, patients who claimed they were sane were thought to be protecting themselves from a truth too shattering to bear. In more recent years, the problem has been reframed as a cognitive deficit intrinsic to the disease. "It has nothing to do with willfulness—you just don't have the capacity to know," Xavier Amador, an adjunct professor of psychology at Columbia University's Teachers College, said. Amador is the author of the most widely used test for measuring insight, the Scale to Assess Unawareness of Mental Disorder, which asks patients why they think their judgments or perceptions have changed. Although researchers haven't uncovered distinct neurological anomalies linked to lack of insight, Amador and other scholars have adopted the term "anosognosia," which more typically describes patients with brain damage who lose the use of limbs or senses yet cannot acknowledge the existence of their new disabilities. Those who go blind because of lesions in their visual cortex, for instance, insist that they can still see, and tell fanciful stories to explain why they are walking into furniture.

Anosognosia was introduced as a synonym for "poor insight" in the most recent edition of the Diagnostic and Statistical Manual of Mental Disorders, but the concept remains slippery, since the phenomenon it describes is essentially social: the extent to which a patient agrees with her doctor's interpretation. For Linda, the validity of her diagnosis was the subtext of nearly all her encounters with her psychiatrists, whose attempts to teach her that parts of her personality could be "construed as a mental illness," as she described it, only alienated her. She wrote to a friend that she was using her hospital stay as an opportunity "merely to prove that I don't have a mental illness (and never did)."

LINDA HAD ALWAYS BEEN FOND of farmhouses. She grew up on Long Island and took pride in her family's sprawling vegetable garden. "My childhood was a good one, with loving and supportive parents who believed in doing things as a family," she wrote in an application to an assisted-housing program. She had a large circle of friends and excelled in school with little effort. "She was bubbly and exuded competence," an old friend, Holliday Kane Rayfield, who is now a psychiatrist, said. Linda's family thought she would become a professor, but she never settled on a professional career. Kathleen White, her closest friend from college, said that her "dream was to find a guy with a sense of humor, have kids, and live on a farm."

Linda got married in 1985, and gave birth to her daughter, Caitlin, five months later. But she complained about her husband's temper, and after her brief marriage ended she struggled to support Caitlin on her own. She worked long hours at a Chinese restaurant in Rochester, New Hampshire, and on her days off she and Caitlin visited museums in Boston or went on camping trips or took aimless drives through the state. Caitlin, now a twenty-five-year-old photo technician at Walmart, told me, "We were each other's world." It wasn't until her mother quit her waitressing job in order to evade the "Chinese Mafia" that Caitlin, who was in seventh grade, began to doubt her mother's judgment. In 1999, in a purple Dodge Dart, the two fled the state, heading toward Canada. Caitlin, too, was terrified of being captured. "I figured I was collateral damage," she said. Linda called friends on the way but lied about her location, because she suspected that she was talking to spies. While her mother used pay phones at gas stations, Caitlin waited in the car. "At some point, I just thought to myself, I know better than this," she said.

When they returned, a little more than a week later, after Linda's fears had subsided, Linda's sister, Joan Bishop, and their parents tried unsuccessfully to persuade Linda to see a doctor. Soon, she disappeared again. She went to Concord, the state capital, to inform authorities that the government was behind John F. Kennedy, Jr.'s plane crash, and then wandered alone through the state for several days, feeling as if she had "ingested some sort of poison or drug without knowing it." Caitlin moved in with her paternal grand-mother and stayed there even when Linda came home. Linda finally checked into a hospital in Dover, New Hampshire, where she was given a diagnosis of schizoaffective disorder, and began taking Zy-prexa, an antipsychotic, and lithium, a mood stabilizer. (Her diag-nosis shifted between bipolar and schizoaffective—a mixture of schizophrenia and a mood disorder—depending on the doctor.) Psychotic disorders typically begin in early adulthood, but it is not uncommon for them to develop later in life, particularly after peri-ods of stress or isolation. Linda sobbed for the first few days, and talked about how betrayed she felt by those who were scheming against her. By the fourth day, though, her psychiatrist wrote, "She now has insight into the fact that these are paranoid delusions, and a part of her is able to say that maybe some of these things didn't happen, perhaps some of the people she felt were plotting against her really weren't." Ten days later, she was released.

It was the beginning of a persistent and common cycle. With each hospitalization, Linda was educated about her illness and the need for medication. This is the standard approach for increasing insight, but it does not account for the fact that people's beliefs, even those which are wildly false, shape their identities. If a person goes from being a political martyr to a mental patient in just a few days—the sign of a successful hospital stay, by most standards—her life may begin to feel banal and useless. Insight is correlated with fewer hos-pital readmissions, better performance at work, and more social contacts, but it is also linked with lower self-esteem and depression. People recovering from psychotic episodes rarely receive extensive talk therapy, because insurance companies place strict limits on the number of sessions allowed and because for years psychiatrists have assumed that psychotic patients are unable to reflect meaningfully on their lives. (Eugen Bleuler, the psychiatrist who coined the term "schizophrenia," said that after years of talking to his patients he found them stranger than the birds in his garden.)

With medication as her only form of treatment, Linda was unable to modify her self-image to accommodate the facts of her illness. When psychotic, she saw herself as the heroine in a tale of terrible injustice, a role that gave her confidence and purpose. After the World Trade Center attacks, in 2001, she moved to New York City, because she felt she had been called to offer her help. A December, 2001, article in the New York *Post*, "Homeless 'Angel' a Blessing at Ground Zero," described how Linda patrolled the perimeter of the site, waving an American flag, greeting visitors, and giving impromptu tours. "Angels come to earth in disguises—some come as paupers," a construction worker was quoted as saying. A man identified as a 9/11 victim said that "God rested on her shoulder." Linda thanked workers at the site for their efforts and talked to tourists about what they were witnessing. "I try to help people understand the enormity," she told the reporter. She dubbed herself the head of Hell's Chamber of Commerce.

FOR THE NEXT FEW YEARS, Linda wandered: she lived on the streets, in homeless shelters, and in her sister's house, on the condition that she take medication. Joan, who works as the director of education for the New Hampshire Supreme Court, has the same warm, jovial manner as her sister, and the two spent much of their free time together, though Linda's goal was to find her own home. In 2003, she entered a supported-housing program in Manchester, New Hampshire, and told her caseworker that she wanted to "live like an adult again." She was upset that her illness had alienated her daughter and friends. Joan told me that "Linda would talk analytically about how it had felt to be delusional. It wasn't a matter of imagining. It wasn't as if she felt she was being chased by government agents. In her mind, they were as real as I am right now."

In the summer of 2004, Caitlin, by then a senior in high school, decided to move back in with her mother for the first time in five years. Their new apartment, in Rochester, New Hampshire, became the preferred hangout spot for her friends. "She was the cool mom," Jessica Jamriska, a close friend of Caitlin's, said. "She had stopped talking about the government, except maybe if there was an election. And the only reason she quoted the Bible is if we were having intellectual debates about, you know, whether it's a book of morals or not."

Linda enjoyed cooking large meals for Caitlin's friends, but over time the stories she told at the dinner table became harder to follow. "At first, we just thought, O.K., it's normal to have some fantasies and dreams," Jessica said. "She would talk a lot about some dude she loved who was going to make everything all right, and we weren't even sure he existed."

Caitlin and Joan urged Linda to take her medication, but she said that she felt perfectly fine and complained that the drugs made her lethargic and caused her to gain weight. (Linda's parents, who had encouraged her to follow her doctors' advice, had died, both of them from cancer, in 2003 and 2004.) Caitlin and two of her friends finally decided to make an audiotape of Linda ranting. "We wanted to have proof, to say, look, this is objectively crazy, and someone needs to help her," Caitlin said. They recorded her talking about how children should be armed with AK-47s, and called the police, but Caitlin said that their complaint was never taken seriously. In February, 2005, Linda's car flipped on its side on Rochester's main street. When the police arrived, they smelled alcohol on her breath. She said that she had purposely caused the accident, to prove "that police officers are 'illegal.'"

Although it was a relatively minor offense (her alcohol level was below the legal threshold), Linda refused to pay the five-hundred-dollar bail, so she was sent to the Strafford County House of Corrections, in March. (Nationally, a quarter of jail inmates meet the criteria for a psychotic disorder.) After her first arrest, Linda threw a cup of urine at a corrections officer and struck a man with a broomstick. Joan wrote to the police department's prosecuting attorney, explaining that before her illness Linda had never been "violent or aggressive towards anyone or anything." She said that the family hadn't been able to get Linda into psychiatric treatment, and asked the attorney to help.

Joan's request led to competency evaluations, and, as Linda waited in jail for the results, she moved even farther away from the life she had led before her illness. She considered herself a "people person"— she made Christmas cards for other inmates out of lunch bags and magazine ads, sealed with grape jelly—but she found herself isolated from all the people with whom she had once been close. She wrote Caitlin long letters with tips about what cosmetics to wear, how to get a job, shop for bargains, lose weight, make apple pie, and avoid the presence of people who belong in Hell, but Caitlin stopped responding.

After a year and a half in jail, Linda was deemed incompetent to stand trial and was transferred to New Hampshire Hospital for a commitment term of up to three years. She was humiliated by the idea of anyone evaluating her competence and wrote to Caitlin, "My constitutional rights have been ignored, trampled on and violated due to your Aunt Joan."

NEW HAMPSHIRE HOSPITAL WAS ESTABLISHED, in 1842, as a kind of utopian community, a reprieve from the disorder of the outside world. The hospital's early leaders tried to help patients regain their common sense—in the first year, more than a quarter of admitted patients suffered from an "overindulgence in religious thoughts," with several claiming to be prophets—by immersing them in a model society. The hospital was situated on a hundred and seventeen acres, and patients lived in a stately, red-brick Colonial building with a steeple and a tiered white porch, surrounded by trees. They farmed, gardened, and cooked together; there was a golf course, an orchestra, a monthly newspaper, dances, and boating on the hospital's pond. In 1866, the hospital's superintendent described psychosis as a "waking dream, which, if not broken in upon, works mischief to the brain," and wrote that the goal of treatment was to "interfere with this world of self—scatter its creations and fancies and people it with objects and thoughts foreign to its own."

As the patient population expanded, though, the hospital couldn't maintain its early idealism. Psychiatrists no longer had time for the benevolent form of care known as "moral treatment." As of 1936, the hospital had sterilized a hundred and fifty-five patients, and later it began experimenting with newfangled remedies, like electroconvulsive therapy and insulin-induced comas; the shock of such procedures, it was thought, might clear patients' minds. By the nineteen-fifties, the hospital's population had swelled to twenty-seven hundred patients, and doctors were less concerned with creating a sense of community than with maintaining security. Patients spent so many years in the hospital that they no longer knew how to leave it. (The institution has two graveyards for people who died in its care.) The hospital's crowded wards resembled those studied in Erving Goffman's 1961 book, *Asylums*, which showed how, through years of institutional life, people lost their identities and learned to be perfect mental patients—dull, unmotivated, and helpless.

The idea that mental illnesses were exacerbated, even caused, by the measures designed to treat them was elaborated by many scholars throughout the sixties. Thomas Szasz, a psychiatrist and prolific author, described mental illness as a "myth," a "metaphor." The psychiatrist R. D. Laing called it a "perfectly rational adjustment to an insane world." In 1963, President Kennedy (whose sister Rosemary had received a lobotomy which left her barely able to speak) passed the Community Mental Health Centers Act, which called for psychiatric asylums to be replaced by a more humane network of behavioral-health centers and halfway homes. His "bold new approach," as he called it, was plausible because of the recent development of antipsychotic drugs, which seemed to promise a quick cure. In the years that followed, civil-rights lawyers and activists won a series of court cases that made it increasingly difficult for patients to be treated without their consent. In 1975, the Supreme Court ruled that the state may not "fence in the harmless mentally ill." Four years later, in Rogers v. Okin, a federal district court decided that involuntary medication was unconstitutional, a form of "mind control." The court maintained that "the right to produce a thought—or refuse to do so—is as important as the right protected in Roe v. Wade to give birth or abort."

Deinstitutionalization was a nationwide social experiment that did not go as planned. Overgrown hospitals were shut down or emptied, but many fewer community centers were opened than had been proposed. Resources steadily declined; in just the past three years, $2.2 billion has been cut from state mental-health budgets. "Wishing that mental illness would not exist has led our policymakers to shape a health-care system as if it did not exist," Paul Appelbaum said in his 2002 inaugural address as president of the American Psychiatric Association. Today, there are three times as many mentally ill people in jails as in hospitals. Others end up on the streets. A paper in the *American Journal of Psychiatry*, which examined the records of patients in San Diego's public mental-health system, found that one in five individuals with a diagnosis of schizophrenia is homeless in a given year.

New Hampshire Hospital, which now has only a hundred and fifty-eight beds, admits people who have been sent from jail or who pose a danger to themselves or others. Often, people arrive at the emergency room, with concerned relatives and friends, but they are turned away, because they are not an imminent threat. "Clinically, it's a shame," Alexander de Nesnera, the hospital's associate medical director, told me. "These are people who may be making choices they would never have made when they were healthy. But then there's the civil-libertarian argument: Who are we to say that they don't have the right to change their opinions?"

Freedom often ends up looking a lot like abandonment. Tanya Marie Luhrmann, a Stanford anthropologist, told me that "there is something deeply American about the force of our insistence that you should be able to ride it out on your own." Luhrmann has followed mentally ill women in Chicago through what is known as the "institutional circuit"—the shelters, halfway homes, emergency rooms, and jails that have taken the place of mental asylums. Many of the women refused assisted housing, because to gain eligibility they had to identify themselves as mentally ill. They would not "formulate the sentence that psychiatrists call 'insight,'" Luhrmann said. "'I have a mental illness, these are my symptoms, and I know they are not real'—the whole biomedical model. To ask for this kind of help is to be aware that you cannot trust what you know."

LINDA READILY ACKNOWLEDGED THAT HER life had gone awry, but she insisted that her diminished circumstances had nothing to do with being "crazy." After reading a booklet on domestic violence, she concluded that her sister was trying to abuse her by convincing others that she was ill and stealing her inheritance. "This will make a great book—a N.Y. Times best seller," she wrote to a friend from the hospital.

Three months into Linda's stay, the hospital filed a petition to make her sister, Joan, her legal guardian, with the authority to force her to take medication. The hospital had to prove beyond a reasonable doubt—the same threshold used in criminal trials—that Linda was incapable of making her own decisions. At the hearing, Linda told the judge that her only problem was that she was "permanently pissed off." "I have a huge amount of medical knowledge," she said. She pointed out that she took daily vitamins, wore anti-embolism stockings, had recently agreed to a mammogram, passed a first-aid course when she was sixteen, and had an uncle who was a podiatrist. To the charge that she was manic, she said, "I've always been like this. Okay? I'm a waitress, I have a lot of energy." The judge found that the evidence of Linda's incapacity—her refusal to accept why she was a patient—did not meet the burden of proof. Linda had become a person that both her sister and her daughter hardly recognized, but the court cannot deprive an individual of her legal rights just because her personality has changed. "She wasn't screaming, she wasn't talking to the ceiling," Joan said.

After the guardianship hearing, which Linda's psychiatrists viewed as their one chance to medicate her, they began talking about her release, though hospital staff continued to express concern about Linda's belief in "the 'plot.'" The psychiatrist who gave a second opinion on her discharge predicted that Linda would "probably get into further altercations with the police" and end up back at the hospital; then there would be more evidence to apply again for a legal guardian. A third of patients discharged from New Hampshire Hospital are readmitted. "Her pattern," he wrote, "is to attract attention to the police before situations become severely dangerous." He seemed to feel that only law-enforcement officials had the power to lead her to treatment.

Linda had spent time away from the hospital on her daily "community passes"—she usually sat in a square in Concord, crocheting and people-watching—but she always returned before dinner was served. She was an active participant in support groups called Inner Strength, Loss & Recovery, and Lifestyle Choices, but she avoided meetings that had a therapy component. According to worksheets she filled out at one group, "freedom" was her only long-term goal.

Her short-term goals included "get clothes," "depart from evil," "put pressure on the guilty," and "laugh more."

A hospital assistant who checked on patients every fifteen minutes to see "if they're breathing" developed a friendship with Linda and encouraged her to think seriously about what she would do after her discharge. But Linda cut off their conversations. "I remember her walking down the hall and she turned around and she said, 'You're putting your values on my life,'" the assistant later explained. "That was my moment. That was, like, I have to back off."

Linda's only plan for supporting herself was to sell some mittens and doilies she had crocheted. Although she had complained of the indignity of being homeless, she didn't authorize the hospital to share her records with a free transitional-housing service, because when she reviewed the paperwork she saw her diagnosis. "I refuse to sign anything that says I am mentally ill," she told her social worker. Instead, she left the hospital with only pocket change, no access to a bank account, and not a single person aware of where she was going.

BECAUSE OF PATIENT-PRIVACY LAWS, LINDA'S doctors never informed Joan or Caitlin that she had been released, and they could not tell them about her condition during her hospitalization, either. Joan sent money and dropped off clothes, but Linda would not see her, and Joan began to feel as though she were only getting in the way. In 2006, she joined the National Alliance of Mental Illness, an organization with twelve hundred chapters created by family members (known as "NAMI mommies") who felt excluded from the medical decisions made by their loved ones. When the organization was founded, in 1979, psychosis was seen as a reaction to a dysfunctional family. According to a prominent theory of the time, the root of schizophrenia was a "schizophrenogenic mother," incapable of communicating her love.

With the rise of brain-imaging scans, which draw links between disturbed behavior and structures in the brain, this line of argument has fallen out of favor. Many psychiatrists and activists argue that current treatment laws are based on disproven theories of mental illness, and patients are given too much freedom to decide what is in their own best interests. The psychiatrist Edwin Fuller Torrey, whose sister had schizophrenia, and who was an early adviser to NAMI, told me that "to keep talking about civil liberty is illogical. Patients are anything but free when they're at the beck and call of their own delusions." In the past thirteen years, Torrey's organization, the Treatment Advocacy Center, in Arlington, Virginia, has lobbied nineteen states to pass or strengthen outpatient commitment laws, which require people to take medication after they've been discharged. (New Hampshire Hospital uses a model called conditional discharge, but in Linda's case, since she would not discuss treatment at all, her psychiatrist decided that engaging in the process would be an "exercise in futility.") In arguing for a more stringent approach to treatment, Torrey, who is also the executive director of the Stanley Medical Research Institute, which has a collection of five hundred and sixty-three brains used for studying the causes of severe mental illnesses, frequently cites research on the significance of anosognosia. He calls it the "single largest reason why individuals with schizophrenia and bipolar do not take their medications." A Treatment Advocacy Center briefing paper on anosognosia features a quotation from a Jacobean play, *The Honest Whore*, by Thomas Dekker: "That proves you mad, because you know it not."

Yet the notion that denying an illness is proof of its existence is a dangerous one, since there are many valid reasons that people choose to refuse treatment, including the stigma of having a mental illness and the disabling side effects of medication. Elyn Saks, a professor of law, psychology, psychiatry, and the behavioral sciences at the University of Southern California, who has schizophrenia, said, "Alleviating suffering is not a legitimate reason for taking away people's rights." When Saks was a law student at Yale, she was restrained and medicated against her wishes; she calls it one of the most degrading experiences of her life. She argues for a greater use of advance directives, which function as psychiatry's version of the living will. In the past two decades, twenty-six states have passed laws that allow people to specify what kind of medical care they wish to receive if they lose their grasp on reality. Some directives include a "Ulysses contract"; just as Ulysses instructed that he be bound to the mast of his ship so that he would not be lured by the Sirens, some people insist in advance that they be medicated or hospitalized even as they beg to be released. At the other extreme, people use directives to reject any future treatment at all.

In her book *Refusing Care*, Saks calls the method "self-paternalism," and argues that there are few other scenarios in which psychiatrists should forcibly impose treatment that intrudes on the privacy of people's own minds. A widely cited justification for compulsory treatment is the "thank-you theory," which assumes that patients will retroactively agree that the intervention was in their best interests. But only about half of patients who have been involuntarily hospitalized subsequently say that they needed the treatment. "We should not be in the business of choosing selves," Saks writes. It's impossible to determine whether a mental illness has altered someone's preferences, or whether that person has simply changed.

The concept of anosognosia is held in disdain by what is known as the "c/s/x movement," composed of consumers, survivors, and ex-patients, who argue that biological explanations for mental disorders have been uncritically embraced. They maintain that one of the reasons so many patients lack "insight" is that they are asked to accept a model of illness that doesn't resonate with their own experiences. Although psychotic disorders have a strong genetic component, popular opinion has swung so far from the logic of the psychoanalytic era, when mothers were blamed for their children's illnesses, that psychosis is often described as if it were inevitable and contextless—a stroke of bad luck. Daniel Fisher, a psychiatrist with a doctorate in biochemistry, who is the executive director of the National Empowerment Center, run by former patients, said that "if you accept the idea that you have this random brain disease, then suddenly your sense of self, your ability to make judgments, and the most fundamental elements of your personality are biologically determined, and this just leads to a sense of meaninglessness and hopelessness."

WHEN LINDA ARRIVED AT THE house on Mountain Road, her mood oscillated between despair and exhilaration at her sudden freedom. For the first time in years, she saw the potential to start her life anew, on her own terms. She finally had her own home, a plot of land, and no one telling her what to do.

During her illness, the person with whom Linda felt closest was a man named Steve Shaclumis, whom she considered the love of her life. Steve met Linda a few times in 1998—"she was a nice waitress who seemed lonely," he told me—but, beyond calling the jail in 2005 and asking it to block Linda's letters, he'd had no contact since. (He was told that inmates could write to whomever they pleased.) While in jail, Linda told Caitlin to design bridesmaid dresses for her and her friends, because she and Steve were getting married. "He wants a big church wedding," she wrote, "which is fine with me."

Linda had spoken of Steve in rapturous tones at the beginning of her illness, but she was self-conscious enough then to recognize her fantasy. She even wrote a letter to a friend explaining that "the process I am going through is not about a relationship with Steve, it is about my relationship with myself." But by the time Linda got to New Hampshire Hospital she considered herself Steve's wife. Now, at the house on Mountain Road, she imagined their domestic routine—they would eat homemade dinners and watch the sunset, holding hands, after "finishing with the needs + business of the day"—and searched the attic for clothes so she would look attractive if Steve arrived. All she could find was a hat.

A plan took shape. Steve would rescue her sometime near Advent, the beginning of the new liturgical year. After gazing at the sky and seeing a cluster of clouds forming the number 4, Linda determined that he'd arrive by December 4th. Using long division and multiplication tables that fill the margins of the last page of her notebook, she calculated that she could survive the "Attic + Apples chapter of this book" if she limited herself to twelve apples per day.

Having spent two years in institutional settings with precisely timed meals and activities, Linda had got into the habit of checking her watch, and she began many of her diary entries, up to four a day, by noting the time. The "high point" of each day was the moment when she crossed off another date on her makeshift calendar, which she usually did around four o'clock. She discovered new books in the attic and under a couch—*What the Bible Is All About, Medical Self-Help Training, Webster's New World Dictionary, Can America Survive?*—but the days dragged. It was difficult to do anything but think about food. She composed long lists of groceries, both a budget list and a "wish list" (vermicelli, peperoncini, V-8, squid), and twenty-five vegetables she intended to grow. She imagined how she'd remodel the kitchen so that she could build a smokehouse like the one in *Little House in the Big Woods*.

What the Bible Is All About was a source of inspiration, and she transcribed a number of verses about the glory of following God's path. Psychosis commonly coincides with religious reveries, and the longer Linda was off her medication the more everyday occurrences seemed to be laced with hidden connections and symbols. Faith in the Lord's plan for her became essential as the days shortened and the temperature dropped. Her neck ached, because she spent about sixteen hours a day curled up under blankets on a mattress, in order to stay warm. So much hair fell out each time she combed it that she realized she might need a temporary wig. But it was "nothing a little fresh air, sunshine, exercise, good food + love won't remedy in a short period of time."

Linda looked out for Steve's white Chevy truck, but by the beginning of December her conviction began to waver. "So maybe the fact that I haven't seen him is a good sign?" she wrote. "I just hope God does want us to be together—everything seems to assure that—but who knows how it all fits. Certainly my death at this point does not seem beneficial to God's plans as perceived by me."

On December 4th, the sight of Christmas lights on a house down the road made her break into tears. She and Caitlin had always enjoyed decorating their tree to excess, and she hadn't seen a lit-up house in two years. By afternoon, Steve still hadn't arrived. Linda was so desperate that she contemplated walking to a neighbor's house and calling a domestic-violence center. But she worried that Satan's workers could be waiting for her. "Dear God," she wrote. "Please save me. I'm trying but don't know what to do. Amen."

The next day, she ate the last of her apples.

ON THE MORNING OF DECEMBER 5th, Linda started a new journal, a spiral-bound composition notebook with full-sized pages that made it easier for her to write while sitting in the "hot seat," a chair she had placed over the only working heating vent in the house. Although the previous night had been "tough," she felt calm and hopeful as she watched the sun rise. The concerns of everyday living briefly receded, and the story she was writing, a tale of love and redemption, became her reality again. "I love my husband so much—and he has done so much to keep me going and give me hope," she wrote. "I can't leave him or give up." She tried reflecting on the positive aspects of her situation—she no longer had to deal with the "babble or yammering" or lies of the hospital—but there was still the problem of her "meal plan." The following day, she wrote, "Facing death by starvation was horrifying and traumatic—and took quite awhile to adjust and consider the whole situation rationally and spiritually." She decided that she had to leave the house. "Before when things needed to change (my location) I always knew—but usually there was an exterior human impetus—however if I stay here—I will die." She planned to hitchhike to the home of an elderly couple, friends of her parents, who lived a few miles away. She would get food, take a shower, and do laundry. From there, she would head to the supported-housing program in Manchester, where she had lived in 2003, the year that her sister thought she had finally recovered.

But walking to the kitchen made her so dizzy that her vision failed. She couldn't imagine how she would get to the road, which was about twenty yards away. She had never doubted her decision to live in the house, and it didn't seem right to abandon faith at such a late stage. "So I'll wait," she wrote, "and continue to pray since God knows where I am."

Linda had never intended to "be clinging to life like some idiot," she wrote, but there was also something willful about her withdrawal from human contact. She knew that if she reëntered society people would be "coding me 'disabled.'" After praying out loud one night, she even questioned whether God "has given me a good brain to figure out what to do." She seemed to accept the premises of two conflicting realities, a phenomenon known as "double bookkeeping," in which psychotic patients who are able to distinguish reality from fantasy go on living and believing in both. Even the patient convinced that he is Christ will abide by hospital curfew and take out the trash. People rarely lack insight in an absolute sense. Solitude allowed Linda's delusions to flourish, and at times she seemed aware of this; she kept away from anyone who could challenge her interpretation of the world.

WITH THE WINTER WEATHER, LINDA'S daily chores required less movement. She snapped icicles off the porch or scooped up bowls of snow and melted them on the heat register, which took about an hour. The task of "harvesting" water was the central activity of her days. She took care to move slowly, since she had already fainted once and had fallen in the kitchen. She conserved energy by lying still on the living-room floor near the register, absorbing its "radiant warmth." To keep her mind active, she forced herself to read the dictionary, but she never got past the letter "K."

On Christmas, Linda reached out the window to get snow for what she predicted would be the last time. Most of the entries that follow are only a few words long, written in faint, uneven letters, because she wrote lying down. Her final note was on January 13, 2008, and it contains nothing more than the day of the week, Sunday. It had been thirty-nine days since she ate her last apple.

On the first page of the notebook, Linda left a note addressed to "whomever finds my body," in which she explained that her death would be the result of domestic violence: "I talked with and wrote to many people in position of authority about this—but no one helped me." She asked to be buried in the cemetery of the town where she had lived when Caitlin was a child and ended the letter with a request, underlined in broad strokes, "Jesus take me home."

JOAN BISHOP DISCOVERED THAT HER sister had been discharged from the hospital when a Christmas card she sent to Linda was returned in the mail three months later. The envelope contained a slip of paper with a new address listed for Linda, but the location didn't exist. After all her attempts to get her sister treatment, Joan was furious that the hospital had "let her walk off the face of the earth." "I work for the justice system; I believe in the justice system," she said. "I still don't understand how you have the right to rot away."

Joan and Caitlin sued the hospital for failing to properly plan Linda's discharge, relying on the police as her only safety net. "Ask the daughter!" Caitlin said. "I could have told them that every time the structure is gone she goes straight down the tubes." The attorneys representing the hospital argued that Linda's doctors had a legal duty to allow her the right to live in the least restrictive setting that her disability allowed. They ended up arguing the case that Linda had been making all along: she wasn't that ill. In the written mediation statement, the attorneys maintained that Linda was making a reasoned decision to pursue an alternative life style. As evidence, they pointed to lucid passages in the journal found at the house after her death: the pleasure she took in identifying birds and commenting on the shapes of clouds; the fact that she realized her unkempt appearance might bring her to the attention of authorities who could hospitalize her again.

Caitlin read the journal, and was hurt that she appeared in it only once, when Linda wrote about a dream in which Caitlin was calling out for her. "I'll never understand that, because my mom was always thinking of me," she told me. Although the suit was settled in mediation for a small sum, Caitlin seemed ambivalent about whether she could blame the doctors. She holds on to an image of a mother who was obstinate and purposeful, rather than overwhelmed by an illness. "The woman who wrote that journal—that was not my mother," she said.

Linda's body was found in early May, when a man interested in buying the house on Mountain Road peered through a window. He contacted the police, who called the owners of the house, a brother and sister who had inherited the property from their parents. One of the owners had checked on the house around Christmas, but she never went inside, because the driveway was covered in snow and there were no footprints. When Caitlin learned of her mother's death, she responded as if she had been waiting for the news for a long time. "My mom made a choice—she could have walked out of that house," she said. "But she wouldn't give up her freedom. She could never let go of that person she always wanted to be."

About the Contributors

RACHEL AVIV's writing has appeared in the *New Yorker, London Review of Books*, the *New York Times Magazine*, and *The Best American Non-Required Reading*. In 2012 she received the American Psychoanalytic Association Award for Excellence in Journalism.

"One of the reasons I was drawn to Linda's story," she explains, "was because she was able to describe the experience of psychosis in her own words. Her diary showed her delusional worldview gradually becoming essential to her sense of who she was. In that context, I found questions about the ethics of forced treatment particularly difficult, since it requires balancing two essential values—freedom on the one hand, mental health on the other—and neither can be clearly defined."

JENNIFER COUZIN-FRANKEL is a staff writer for *Science* magazine, where she covers various topics in medicine, basic biology, and the scientific community. Her work has been published in *Wired*, *U.S. News & World Report*, and the *Washington Post*, among other publications. This is her fourth appearance in *The Best American Science Writing*; she has also won the Evert Clark/Seth Payne Award, given annually to a young science journalist. She grew up in Toronto and lives with her husband and two children in Philadelphia.

"This story began for me more than eight years ago, when I wrote about a dispute between two prominent researchers who study aging," she recalls. "That piece appeared in the 2005 volume of *Best American Science Writing*, but both the personalities and the science stayed with me. The researchers were probing a fascinating mystery: why slashing calories makes animals live longer.

"I often felt there was a sequel to that first piece, and as the years passed I watched the science unfold. The story published here focuses on new research that's challenging early work done mainly in a single lab at MIT—with the gauntlet thrown by some scientists who worked there as grad students and postdocs. I was especially fascinated by how people can interpret the same scientific results differently. And also by young scientists who challenge their mentor's work, whatever the consequences."

DAVID DOBBS writes for the *Atlantic*, the *New York Times Magazine*, *National Geographic*, *Nature*, Wired.com, and other publications. He is currently writing a book with the working title *The Orchid and the Dandelion* (Crown) that explores the notion that some of the genes underlying our most vexing mood and behavior problems—melancholy, madness, murder—may also generate some of our greatest strengths, accomplishments, and contentment. He is the author of three previous books, most recently *Reef Madness*, and of the #1 Kindle Single "My Mother's Lover." He keeps his blog, *Neuron Culture*, at Wired.com.

"My favorite response to this story—to almost any story I've done—came from a fifteen-year-old," he notes. "The kid was with his father, an acquaintance of mine in our tiny Vermont town, when I bumped into them outside the hardware store. 'Thanks,' he said. 'I don't even understand teenagers. But when I read that I thought, "That's me." It was like you were in my head.' I've never had a better reader.

"My son, by the way, who did 113 to generate my lede, is splendid. He now takes his risks not by driving, but through college, where he's training for a career in journalism."

JOSH FISCHMAN is the senior science writer for the *Chronicle of Higher Education* and previously supervised science and technology coverage as a senior editor. He joined the publication in 2007. He has written about a reinvented internet, cognitive tutors in class, the controversy over engineered bird flu viruses, the aging brain, and water on Mars. Previously he was a senior writer and editor at *U.S News & World Report*, editor in chief at *Earth*, deputy news editor at *Science*, and a senior editor at *Discover*. He has won the Blakeslee Award for excellence in medical reporting, and has been a finalist for National Magazine Awards. He is a frequent contributor to national magazines such as *National Geographic* and author of the leading medical education guidebook, *The U.S. News & World Report Ultimate Guide to Medical Schools* (Sourcebooks, 2006).

"When I heard Adrian Raine say, at a scientific meeting in early 2011, that violent behavior in children could be predicted even before the children commit antisocial acts," he recalls, "it conjured up visions of a controlled society with loss of freedom, as well as astonishment that science has come so far. The high-tech vision contrasted sharply with Raine's lab when I visited it a month or so later. It was crammed into a Philadelphia row house owned by the University of Pennsylvania, and the small rooms housed little equipment beyond a souped-up lie detector. But these humble surroundings are producing cutting-edge work that, Raine says frankly, is going to force society to confront very uncomfortable questions: What kind of confidence in this science do we need before we intervene? Will intervention be therapeutic, or focused on containment of dangerous persons? How can we do any of this in a legal system that protects people who have not committed a crime? What struck me about Raine was his willingness to confront these questions using his own two children as examples. He says if they were at risk for being criminals, he'd want to know. If the test that revealed that information meant he would lose custody of them, he's not so sure."

TIM FOLGER has been writing about science for more than twenty years. His work has appeared in *Discover, National Geographic, Scientific American, On Earth, Science,* and other magazines. In 2007 he won the American Institute of Physics Science Writing Award. He lives with his wife in northern New Mexico.

"While working on this story," he says of "Waiting for the Higgs," "it was difficult to avoid thinking that our country's leaders have made—and continue to make—some ruinous decisions when it comes to science. Nineteen years ago, Congress canceled the funding for a successor to the Tevatron, the National Science Foundation today supports fewer research proposals than it did a decade ago, and one of the two leading political parties denies the reality of global warming. We need to change our national priorities."

DOUGLAS FOX is a freelance writer based in Northern California. His stories have appeared in *Discover, Scientific American, Esquire, Popular Mechanics, New Scientist,* the *Christian Science Monitor,* and other publications.

"I first scribbled the idea for this story in the margin of a scientific paper that I was reading—one of Simon Laughlin's papers," he notes. "Realizing that thermodynamics might place a fundamental limit on intelligence was a eureka moment. But at the time I didn't have the confidence to put it into print (journalists, after all, are supposed to report on crazy ideas—not invent them). So for several years, every time I took that paper out, there was my scribbled idea, staring at me. A friend finally convinced me to write the story in 2009."

RIVKA GALCHEN is the author of the award-winning novel *Atmospheric Disturbances*. Her stories and essays have appeared in the *New Yorker, Harper's Magazine*, the *Believer*, and the *New York Times*.

She writes: "A small update on quantum computers: a particle first hypothesized in 1937, called the Majorana fermion—a fermion that is also its own antiparticle—seems now to have been discovered, and it will almost certainly make the building of a quantum computer much easier. Also the case of the eponymous Ettore Majorana, the Italian physicist who disappeared from fascist Italy the year after he hypothesized the particle, has recently been reopened, following the surfacing of photos of a man in Argentina in the 1970s who may have been Majorana."

JEFF GOODELL is a contributing editor at *Rolling Stone* and the author of five books, including *Big Coal: The Dirty Secret Behind America's Energy Future* and *How to Cool the Planet: Geoengineering and the Audacious Quest to Fix Earth's Climate*, which won the 2011 Grantham Prize Award of Special Merit.

"I wrote 'The Fire Next Time' in the weeks following Japan's Fukushima nuclear disaster in 2011," he explains. "The event raised questions about the vulnerability of America's aging fleet of nuclear plants, as well as about the power of the Nuclear Regulatory Commission to oversee those plants and ensure their continued safe operation. Despite the risks that were highlighted in my story, one year later, the NRC is still dragging its feet on reforms, putting the lives of one hundred million Americans who live within fifty miles of a nuclear plant at risk."

DENISE GRADY has been a reporter in the science news department of the *New York Times* since September of 1998. She wrote for the *Times* for several years before that as a freelancer. Grady has written more than seven hundred articles about medicine and biology for the *Times*; has edited two *Times* books, one on women's health and another on alternative medicine; and is the author of *Deadly Invaders*, a book about emerging viruses that was published in October of 2006. Her work for the *Times* has included reporting on health from Sri Lanka, Angola, Tanzania, and Jordan.

"The research in this article was first described by the scientists in two journals," she recalls, "in reports that were so dense and technical it was hard to figure out what was going on. But one of the health editors at the *New York Times*, Mike Mason, said he thought there was a story buried in the data. He turned out to be right. The last I knew (in March of 2012), William Ludwig and Patient 3 were still doing well."

JARON LANIER is a computer scientist, composer, artist, and author who writes on numerous topics, including high-technology business, the social impact of technology, the philosophy of consciousness and information, internet politics, and the future of humanism. In 2010 Lanier was named one of *Time*'s 100 Most Influential People in the World. He has also been named one of *Prospect* and *Foreign Policy* magazines' Top 100 Public Intellectuals in the World, and one of history's three hundred or so greatest inventors in the *Encyclopedia Britannica*. In 2009 Jaron Lanier received a Lifetime Career Award from the IEEE, the preeminent international engineering society. A pioneer in virtual reality (a term he coined), Lanier founded VPL Research, the first company to sell VR products, and led teams creating VR applications for medicine, design, and numerous other fields. He is currently partner architect at Microsoft Research. He is the author of *You Are Not a Gadget: A Manifesto*, which was published in 2010 by Alfred A. Knopf, and is currently working on his next book, *The Fate of Power and the Future of Dignity*, to be published by Free Press in 2013.

JACKSON LEARS is Board of Governors Professor of History at Rutgers University and editor in chief of *Raritan: A Quarterly Review*.

He is the author of several books in cultural history, most recently *Rebirth of a Nation: The Making of Modern America, 1877–1920*. He has been a regular contributor to the *London Review of Books*, the *New Republic*, and the *Nation*, among other publications. In April of 2009, he was elected a fellow in the American Academy of Arts and Sciences.

"For a long time," he states, "I have been troubled by the tendency to turn scientific method into an absolutist metaphysic—an outlook very far from a genuinely experimental frame of mind. The polemics of the New Atheism display this misunderstanding flagrantly, and spread a lot of confusion as a result. This essay gave me an opportunity to clear the air."

ALAN LIGHTMAN is a novelist, essayist, and physicist. He was the first person to receive dual faculty appointments at MIT in science and in the humanities. His literary work has appeared in the *New York Times*, the *Atlantic*, the *New Yorker*, and other publications. Lightman's novel *Einstein's Dreams* was an international bestseller and has been translated into thirty languages. His novel *The Diagnosis* was a finalist for the National Book Award in fiction. His new novel, *Mr g*, the story of creation as told by God, was published in January of 2012. Lightman is also the founding director of the Harpswell Foundation, which works to empower women leaders in Cambodia.

"For me, theoretical physics is the branch of science closest to philosophy and religion," he says. "The subject raises fundamental questions about our existence. Although I retired as a practicing theoretical physicist some years ago, I have tried to keep up with the major developments in the field, and the multiverse is the most interesting and provocative of these."

CHARLES C. MANN is the author, most recently, of *1491*, which won the National Academy of Sciences' Keck Award for best book of the year, and *1493*, which was a *New York Times* bestseller. A correspondent for *Science, Wired,* and the *Atlantic,* he has written for many newspapers and magazines in the U.S., Europe, and Japan. In 2007 the American Association of Geographers named him an Honorary Geographer.

"The people at *National Geographic* suggested this story to me and went to enormous trouble and expense to make it work out," he adds. "If readers enjoy the story, as I hope they will, they should remember that these articles are not produced by writers alone. My thanks to the editors, illustrators, researchers, and designers at the magazine."

LINDA MARSA is a contributing editor for *Discover* and an award-winning journalist who writes about science, medicine, and the environment. A former *Los Angeles Times* staff writer, her work has also appeared in *American Archaeology, Popular Science, Utne Reader, Financial Times, Playboy, Los Angeles, Los Angeles Times Magazine,* and many others. She is the author of *Prescription for Profits* (Scribner), and the upcoming *Fevered: Why a Hotter Planet Will Make Us Sick and How We Can Save Ourselves,* which is due out in the spring of 2013.

"While doing research on climate change," she writes, "I discovered Australia was already being hit hard by global warming. I decided to see 'the canary in the coal mine' for myself and spent nearly a month crisscrossing the vast island continent. What I uncovered was both frightening and inspiring. The land down under has long been whipsawed by extremes in weather, but rising temperatures have increased the incidence and intensity of natural disasters, unleashing record floods, severe decade-long droughts, killer heat waves, and wildfires of unimaginable ferocity. But the hardy, resilient Aussies—who more than lived up to their reputation as incredible hosts—are fighting back and I got a chance to see some of their innovative solutions, which may soon become models for the rest of the world."

P. J. O'ROURKE is a contributing editor at the *Weekly Standard*, the H. L. Mencken Research Fellow at the Cato Institute, and a member of the board of directors of the Space Foundation. He also avoided flunking high school physics only because the high school physics teacher was also the summer-school physics teacher, and an F would mean trying to teach physics to Mr. O'Rourke *again*.

CHARLES PETIT has been writing about science for forty years—twenty-six of them for the *San Francisco Chronicle*, more recently for *U.S. News & World Report*, and currently as a freelancer. He lives in the Bay Area, is a former president of the National Association of Science Writers, and, since 2006, has been chief writer and critic for the MIT Knight Science Journalism Tracker, which may be found at ksjtracker.mit.edu.

"After following and occasionally reporting on the sensational Kepler planet-finding satellite for some time," he says, "I happened upon a session about it during a meeting in Washington early in 2011. I expected no news I could sell. I was merely taking a pleasant break. As soon as a Kepler manager pleaded for help with dumbfounding data having nothing to do with planets, I said to myself, 'Pay dirt!'"

DEVIN POWELL is a staff writer at *Science News* in Washington, D.C. His stories have appeared in *New Scientist* and online for *ScienceNOW*, MSNBC, *ABC News*, *FOX News*, *U.S. News & World Report*, and *Discovery News*. Several newspapers have also published his work, including the *Minneapolis Star Tribune* and the *Knoxville News Sentinel*. He holds a bachelor's degree in biochemistry from Harvard and a master's degree in science writing from Johns Hopkins University.

"When I started 'Moved by Light,' I thought the race it describes had already been won," he notes. "A landmark study in 2010 reported the first relatively large, fairly ordinary object to behave quantum mechanically (a ceramic wafer that both moved and stayed still at the same time). But I soon discovered the finish line was as fuzzy as the quantum world itself. Bigger objects still plodding forward promise to behave more weirdly—helping scientists probe the boundaries between the sensible large scales that humans inhabit and the utter bizarreness that reigns at the smaller scales atoms call home."

EVAN RATLIFF is the cofounder and editor of the *Atavist* (www.ata-vist.com) a hybrid magazine/book digital publisher that produces narrative nonfiction. He is also an award-winning journalist, whose feature writing on science, technology, crime, and terrorism appear in *Wired*, the *New Yorker, National Geographic*, and many other publications.

"The fox farm experiment is one of those science endeavors that enthralls on its first mention," he says. " 'Wait, they turned foxes into something like dogs?' My introduction came through a doctor friend who'd stumbled across it as part of some unrelated research, and mentioned it in passing. I was baffled to discover that no reporters had been there for years; it wasn't even clear at what level the experiment was still operating. When *National Geographic* gave me the opportunity to go investigate, I expected it to be fascinating. I didn't expect that it could also be heartbreaking."

KRISTINA REBELO is an award-winning investigative reporter based in Southern California. She worked for *Sports Illustrated* for nine years and was also published in *People* and *Time*. She has produced reports on government mishap cover-ups for television, including one of the highest-rated segments on the popular series *Hard Copy*. WALT BOGDANICH is a three-time Pulitzer Prize–winning investigative journalist who has worked at the *New York Times* for twelve years. Prior to that, he won a Pulitzer with the *Wall Street Journal* and was an accomplished investigative television journalist.

The series, "Radiation Boom," from which "X-Rays and Unshielded Infants" was a part, won a public service first-place award from the Associated Press Managing Editors.

"This series was the product of three years of really dogged reporting," they explain. "Our investigation found that the surge of radiation treatments and procedures had outpaced clinical skills and quality-assurance capabilities. We were gratified that our reporting got the attention of not only the public but the International Atomic Energy Agency, the Nuclear Regulatory Commission, and the medical establishment.

"Around the world, those in the radiology and health physics fields continue to hold frequent training symposia on radiation safety. And according to an American Society for Therapeutic Radiology and Oncology report, vendors have retooled the manufacturing of radiation-producing machines and software with a goal of reducing the occurrence of errors or malfunctions.

"Perhaps the most surprising—and in some ways most flattering—testament to our stories' impact came when we learned that members of the American Association of Physicists in Medicine now use the *Times* as a verb when discussing quality assurance caveats—as in, 'You better be careful with radiation dosages; you don't want to be *New York Times*-ed.'"

ERIK SOFGE is a journalist who covers science, technology, and culture. He is a contributing editor at *Popular Mechanics*, and coauthor of *Who's Spying On You: The Looming Threat to Your Privacy, Identity, and Family in the Digital Age*, due this fall from Hearst Books. His work has also appeared in *Men's Journal*, *Slate*, and the *Wall Street Journal*. He lives with his wife and daughter in Massachusetts.

"It's inevitable that many—maybe most—people will write off space tourism as yet another way for the ultrawealthy to burn off their descendants' inheritances," he notes. "I certainly felt that way. But, cliché as it sounds, good assignments prove you wrong. In piecing together this educated guess at what private spaceflight will look and feel like, from the g-forces that threaten to club you unconscious, to the neo-mystical experience of watching Earth from orbit, that sense of overprivileged joyriding vanished. This is a brave sect of rich folk, willing to harrow the heavens for the rest of us. That's worth our respect."

GRETCHEN VOGEL is a contributing correspondent for *Science*. She has covered stem cells for the magazine for more than a decade, and she also writes about developmental biology, infectious disease, and European science policy. She lives with her husband and two daughters in Berlin.

"Although stem cells have gotten a huge amount of press (and hype)," she says, "real applications in the clinic are still relatively rare. Breuer's and Shinoka's careful approach impressed me. I also enjoy finding stories that overturn my assumptions. I liked the fact that it wasn't the seeded stem cells that formed the new vessel, but rather the body's own innate repair mechanisms. Indeed, the researchers say the cell-free grafts mentioned at the end of the story are already showing promising results in animals."

STEVEN WEINBERG taught at Columbia, Berkeley, MIT, and Harvard, where he was Higgins Professor of Physics, before joining the University of Texas as Josey Professor of Science in 1982. He has received the Nobel Prize in Physics, the National Medal of Science, and numerous other awards, and has been elected to the U.S. National Academy of Sciences, Britain's Royal Society, and other academies. His articles for general readers have been collected in *Facing Up* and *Lake Views*. He received his A.B. and Ph.D. from Cornell and Princeton and also holds honorary doctoral degrees from sixteen other universities.

"Like many of the articles I have written," he explains, "this essay is based on a talk I gave at a conference devoted to symmetry at the

Technical University of Budapest in August of 2009. It wasn't hard to decide to accept the invitation to speak at this conference, as symmetry has been at the center of much of my work in theoretical physics, and I had always wanted to visit Budapest. I sent the article to the *New York Review of Books*, which has published my essays from time to time; this one appeared in November of 2011."

Permissions

A Note from the Series Editor

Submissions for next year's volume can be sent to:

Jesse Cohen
c/o Editor
The Best American Science Writing 2013
HarperCollins Publishers
10 E. 53rd Street
New York, NY 10022

Please include a brief cover letter; manuscripts will not be returned. Submissions can be made electronically and sent to BASWEditor@gmail.com.